Controversies in Science
and Technology

SCIENCE AND TECHNOLOGY IN SOCIETY

Series Editors

Daniel Lee Kleinman
Jo Handelsman

Advisory Board

Caitilyn Allen
University of Wisconsin–Madison

Wendy Baldwin
University of Kentucky

David Baum
University of Wisconsin–Madison

Patricia Flatley Brennan
University of Wisconsin–Madison

Sandra Harding
University of California–Los Angeles

Cora Marrett
University of Wisconsin System

Catherine Middlecamp
University of Wisconsin–Madison

Kelly Moore
City University of New York

Andrea Tone
McGill University

Controversies in Science and Technology

From Maize to Menopause

Edited by
Daniel Lee Kleinman, Abby J. Kinchy,
and Jo Handelsman

THE UNIVERSITY OF WISCONSIN PRESS

This book was published with the generous support of the Brittingham Fund and of the Evjue Foundation, Inc., the charitable arm of the *Capital Times*.

The University of Wisconsin Press
1930 Monroe Street
Madison, Wisconsin 53711

www.wisc.edu/wisconsinpress/

3 Henrietta Street
London WC2E 8LU, England

Library of Congress Cataloging-in-Publication Data
Kleinman, Daniel Lee.
Controversies in science and technology: from maize to menopause /
Daniel Lee Kleinman, Abby J. Kinchy, and Jo Handelsman.
p. cm.—(Science and technology in society)
Includes bibliographical references and index.
ISBN 0-299-20390-5 (hardcover: alk. paper)—
ISBN 0-299-20394-8 (pbk.: alk. paper)
1. Genetic engineering—Popular works. 2. Antibiotics in agriculture—Popular works.
3. Menopause—Hormone therapy—Popular works. 4. Bioterrorism—Popular works.
5. Science—Social aspects—Popular works. 6. Technology—Social aspects—Popular works.
I. Kinchy, Abby J. II. Handelsman, Jo. III. Title. IV. Series.
QH442.K57 2004
303.48′3—dc22
2004012829

This book is dedicated to the memory of Blossom Handelsman,
who lost the battle with antibiotic-resistant bacteria,
and to
Fred Buttel, who has been a teacher and friend to all three of us.

Contents

Acknowledgments

Producing a volume with nineteen essays on four different topics is a massive undertaking, and we could not have done it alone. First and foremost, we have to thank the authors who contributed to this collection. We asked a very busy group of people to find time to write engaging pieces in a very short time frame, and all did so. In addition, we asked a number of specialists to review sections of the manuscript—again, in short order. The reviewers helped take already engaging and provocative work and make it even better. Next, we received useful advice about topics, authors, and reviewers from our advisory board. We would like to thank all for their assistance.

Beyond advice and writing, this project required financial resources and institutional support. We would like to thank the Brittingham Fund, the Howard Hughes Medical Institute (which supports Handelsman's work), the U.S. Department of Agriculture (which supports Kleinman's work through a Hatch grant), and the University of Wisconsin Press for providing the financial wherewithal that made production of this book possible. In addition, virtually the entire staff of the University of Wisconsin Press helped in some fashion to get this book into print. Most especially, we would like to thank Robert Mandel, director of the Press, for his support of the *Controversies in Science and Technology* and the book series of which it is a part. In addition, we would like to thank Erin Holman for her editorial work on the volume, and Andrea Christofferson, David Herzberg, Sheila Leary, and Gwen Walker for their many contributions to the production and promotion of this book. Polly Kummel did an extraordinary job copy-editing the manuscript and the collection is stronger as a result.

A few additional individuals warrant our thanks. Throughout the process of putting the book together we benefited from Christina Matta's knowledge of the history of science and contemporary science. Regina Vidaver helped behind the scenes in getting this volume into production and in building the book series more generally. In the last days of getting the manuscript out the door, we were assisted by Tracy Schmidt. We also benefited from the administrative assistance of Mischelle Grabner and Kari Straus. Finally, Jack Kloppenburg introduced us (Kleinman and Handelsman) fifteen or more years ago, and without that meeting, this book would not be before you.

Controversies in Science
and Technology

Introduction

From Maize to Menopause

Abby J. Kinchy, Daniel Lee Kleinman, and Jo Handelsman

It is unusual for a day to go by when science- or technology-related matters do not find their way into the news. The press has widely covered the debate about the use of human stem cells in research and medical therapy, and its discussion among policymakers, ethicists, scientists, and friends has been widespread. In the wake of the space shuttle *Columbia* disaster, the future of the U.S. space program has been on the policy table, and this discussion has raised important questions about the value of space science and the role of tax dollars in supporting it. Preliminary evidence suggesting the possibility that life once existed on Mars has prompted deep questions about the universe and our place in it.

While we were putting the final touches on this book, the Bush administration was taken to task by a group of more than sixty of the U.S.'s most prominent scientists for distorting and suppressing scientific findings that contradict agency policies and for undermining the independence of government science advisory panels. In Mendocino, California, voters took science into their own hands by passing an initiative that will prohibit commercial use of genetically engineered crops, and in Germany, a French biotechnology company announced the results of research that, it contends, promises improved treatment for breast cancer. Whether it is food we eat, the medical treatment we or our loved ones receive, or the cell phones on which millions of us log thousands of minutes a year, developments in science and technology affect our lives every day in ways that sometimes are close and tangible and other times more abstract—but always thought-provoking.

This is the inaugural volume of what we expect to be a semiannual collection of essays covering science- and technology-related issues likely to be of wide public interest and of substantial public importance. The book has four sections. First, our contributors explore the growing debate about antibiotic use in livestock and antibiotic resistance. In the second section the writers consider the recently prominent phenomenon of gene flow and the related question of the virtues and drawbacks of genetically modified crops. In the third part of the book the authors discuss hormone replacement therapy, and in the final section contributors write about infectious disease and bioterrorism.

A prominent figure introduces each section with an overview of the topic. These overview essays are designed to provide readers with a broad introduction to the technoscientific matters at stake in the area under examination. Several chapters follow the overview essay, with each chapter providing a very different perspective on the subject. We make no claim to comprehensiveness in our coverage—several volumes could be written about each topic at hand. Instead, we searched widely for reflective and engaging voices on the topics that this book covers. The writers include scientists, historians, journalists, a sociologist, an anthropologist, legal scholars, and a policy analyst.

Our aim is to increase understanding and broaden dialogue. Although there is disagreement about what science literacy means and how widespread technoscientific illiteracy is, no one doubts that because developments in science and technology increasingly affect the day-to-day lives of people throughout the world, providing accessible and thought-provoking perspectives on these issues is a good thing. We hope that after reading these essays readers will be in a better position to think about the issues at stake, to explore matters further, and to participate in some of the central debates of our time.

Overuse of Antibiotics on the Farm

Widespread antibiotic resistance among human pathogens threatens to return us to an era of medicine that many associate with leeches and amputations. With the advent of penicillin, World War II was the first war in which more American soldiers died from bullets than from bacterial infections. The development of penicillin, and the proliferation of antibiotics that followed, transformed human existence, moving infectious disease far down the list of causes of human mortality (Lederberg 2000; Interagency Task Force 1996). During the last few decades of the twentieth century, infectious diseases caused by bacteria were considered treatable, not deadly. Many of us grew up in an era in which an earache, sinus infection, or urinary tract infection meant a trip to the doctor and prescription for an antibiotic—a treatment that was usually effective and left us feeling fine after mere days. Even worldwide health threats such as tuberculosis, which affects one-third of the world's population, were considered manageable; the challenge no longer was finding a treatment but getting the right antibiotics into the hands of the people afflicted. The success of antibiotics in managing infectious disease led many pharmaceutical companies to terminate their antibiotic research programs, replacing them with research on chronic conditions such as diabetes, asthma, and high blood pressure (Projan 2003; Wilson 2002). Drugs that treat these conditions provide a greater financial return on research investments than do antibiotics, which are usually taken for short treatment cycles. These trends, coupled with the power of bacterial evolution, have produced a health crisis in which bacteria may once again present a greater threat than conventional warfare.

The extensive use of antibiotics has caused the spread of antibiotic resistance among bacteria. Some members of any population of bacteria will be naturally resistant to an antibiotic. When they are exposed to the antibiotic, the resistant minority will multiply and eventually outnumber the susceptible bacteria, making it more likely that future hosts will be infected by a pathogen that is resistant to antibiotics. Some pathogens develop resistance to multiple antibiotics, making them difficult to treat despite the arsenal of antibiotics that has developed since the 1940s. The explosion of antibiotic resistance among major pathogens such as *Staphylococcus, Enterococcus,* and *Pseudomonas* are of particular concern to people who are immunocompromised, severely burned or wounded, undergoing chemotherapy, or in hospitals, where exposure to antibiotic-resistant pathogens is high (Clark and Hershberger 2003; Hiramatsu and Okuma 2002; Interagency Task Force 1996; Mehta and Neiderman 2003; Sheldon 2003). The threat is extending to all people, and even common cuts and coughs are causing concern among doctors, parents, and patients because the infections do not respond to a traditional course of antibiotics. There is little controversy about the assertion that medical use of antibiotics contributes to development of antibiotic resistance among human pathogens (Austin and Kristinsson 1999; Conway et al. 2003; Enright 2003; Mackenbach and Looman 1988; Price and Honeybown 2004; Weigel and Clewell 2003). What is far more controversial is the effect of antibiotics in animal husbandry on the development of resistance among human pathogens. In the first section of this volume the authors explore this controversy and the nature of controversy itself.

In the overview Christine Mlot, a bacteriologist–turned–science writer, describes the biosphere as bathed in tetracycline. She cites data showing that half of the one million metric tons of antibiotics in the biosphere originally were used in raising animals for food. Animals are treated for acute infections, but they are also fed low doses of antibiotics to boost their growth, and it is this practice that is so controversial. Mlot reviews the evidence that resistant bacteria, as well as their genes, flow easily between animals and the humans who care for them. She explores the regulatory responses to the issue, as well as alternatives to antibiotic use that might reduce the selective pressure for resistance.

Brian Martin, an applied mathematician and social scientist, places the antibiotic debate in the context of other public controversies. He highlights the features that have typified the debates about fluoridation of public water and nuclear winter—polarization, the distraction from discussion of alternatives, the partisan nature of participants and their special interests, the nature of the attacks, and evidence. Many scientists believe that if we just had sufficient evidence on one side or the other, the controversy would be resolved. But Martin argues that public debate is always about more than just evidence and that evidence is sometimes secondary to political concerns and other factors that can dominate a particular position. The antibiotic debate shows genuine disagreement about the evidence, but the controversy encompasses more than disputed data or a lack of data.

Abigail Salyers, a renowned molecular biologist who has revolutionized our understanding of the movement among bacteria of genes that are resistant to antibiotics, provides insight into molecular events that underpin the spread of genetic resistance. She explores the basis for concern about the risk of antibiotic resistance, the limitations of the data available, and the caveats that affect interpretation of those data. Salyers argues that although food distributors such as McDonald's and Tyson have the autonomy to decide to use antibiotic-free meat without convincing data to support the decision, the U.S. Food and Drug Administration (FDA) and the U.S. Department of Agriculture (USDA) usually develop policies based on solid scientific research. In this case, she claims, the scientific foundation that the agencies are being asked to use is filled with enormous holes. Some gaps are hard to fill. For example, we know little about the frequency of antibiotic resistance among bacteria in the environment before the widespread use of antibiotics. Assessing the effects of antibiotic use without such data is challenging. Salyers advocates evidence-driven policy and a balanced role for scientists in public controversy.

The third chapter is written by Randall Singer, a veterinarian and epidemiologist, who further explores the gaps in our knowledge. Singer examines the veterinary science behind antibiotic use, revealing the potential risks of *not* using these drugs. A pivotal example is the proposed ban on the use of fluoroquinolones in poultry, which has been the subject of public hearings and vigorous debate at the FDA. The fluoroquinolones are used to treat acute lung infections in poultry caused by *E. coli*. Because these drugs are also essential weapons in the antibiotic stockpile for treating life-threatening infections in humans, the vociferous proponents of a complete ban on nonhuman uses are numerous. Singer questions whether banning the use of fluoroquinolones in animals would, in fact, lead to more human exposure to *E. coli* and consequentially more infections, as well as to higher mortality in poultry.

Singer also highlights a common theme in the public debate of scientific issues: People (including scientists) assume causality when only a casual association has been demonstrated. This is true in a number of studies on the effects that antibiotic use in animal husbandry has upon antibiotic resistance in human pathogens. There is often a gaping hole in the data that would connect the antibiotics, resistant bacteria, animals, and humans, but these gaps are often filled in with arguments built on a weak foundation. The final chapter in this section, by Tamar Barlam, a physician and former director of the Project on Antibiotic Resistance at the Center for Science in the Public Interest, echoes Salyers's and Singer's calls for further investigation. But Barlam argues that the absence of complete information should not prevent the development of responsible policy. She painstakingly traces the paths of antibiotic-resistant genes through food and the environment, linking farm use of antibiotics with resistance in human infections. Barlam uses historical examples of harm caused by slow development of public policy to reinforce her exhortation for immediate action in this case. In retrospect, it is easy to see that pasteurizing milk and removing lead from gasoline were for the public good. Will we look back and think the same way about regulating antibiotic use in animal agriculture?

Genetically Modified Crops: Global Issues

In the early 1970s, Stanley Cohen and Herbert Boyer developed recombinant DNA (rDNA), a technique for isolating and making multiple copies of a piece of DNA or an entire gene and for transferring genetic material from one organism and (re)combining it with DNA from another. This technology makes it possible to combine genetic material from two distantly related organisms and thus allows researchers to circumvent the barrier to combining genetic material across species.

In the heady 1970s, scientists and investors imagined that rDNA and related techniques would provide the foundation for a radical transformation of agriculture. Although not in precisely the ways they expected, their vision of transformation was broadly correct. By 2001, farmers had planted nearly half the world soybean crop with genetically engineered seed. Worldwide, according to the Action Group on Erosion, Technology and Concentration (2002), the total area planted in genetically engineered crops expanded from less than two million hectares in 1996 to more than fifty million in 2001.

Tremendous controversy has accompanied the technical advances and the growth in commercial possibilities. Many argue that this new technology will allow us to feed the hungry and reduce our use of chemical pesticides (see chapter 10). Others question whether humans should "play God" by altering the genetic blueprint of life (Howard and Rifkin 1977). Still others worry about the ways in which developments in genetically modified agriculture will tighten the grip of large corporations on agricultural production and further undermine the viability of small-scale farming (Kloppenburg 2004). Finally, some critics are concerned about the potential for environmental disaster—the possibility that genetic material that strengthens the vitality of crop plants will find its way into weeds, making them unbeatable pests or narrowing genetic diversity by contaminating natural crops with genetic material from genetically altered plants (Greenpeace 2004).

The controversy has led to serious and sustained discussion worldwide about how to regulate genetically modified crops. In the United States, surveys register some public discontent, but commercial development has proceeded relatively smoothly. In Europe, by contrast, public reaction has led to serious consideration of highly restrictive regulation and to widespread labeling of genetically modified foods for sale by grocers (Gaskell et al. 2001). Public opposition to genetically modified foods in Europe has been so powerful that suppliers are increasingly unwilling to accept agricultural products from the United States (see chapter 9).

In this section of the book, we hope to enrich discussions by providing basic information and some perspectives often overlooked in dueling headlines and sound bites. The section begins with an overview essay by Allison Snow, a biologist whose work focuses on the environmental effects of genetically engineered crops. Snow begins by describing the process through which the United States regulates genetically modified crops. She then turns her attention to the problem of gene flow—the spread of genetic material through the dispersal of pollen and seeds—focusing on maize from Mexico and the United States. Carefully traversing the existing scientific literature, Snow considers the potential environmental dangers of genetically modified corn currently on the U.S. market. Although further research is certainly appropriate, she says that, so far, environmental problems appear to be minimal. As for Mexico, although research suggests some genetic material has transferred unintentionally from genetically engineered corn to Mexican maize landraces (locally adapted strains of specific species—plants and animals—bred through traditional methods of selection, generally in subsistence agriculture), this is to be expected and is not a major threat to genetic diversity, according to Snow. Still, she concludes that we should proceed with caution, thorough debate, and ongoing research.

In chapter 7, Paul Gepts, an agronomist who studies the mechanisms responsible for crop biodiversity, expresses a need for caution in the use of transgenes (genes artificially inserted into an organism's native genome), given their potential for affecting centers of domestication, geographical locations where plants were initially domesticated and systematic crop production was initiated, like Mexico. Gepts notes that little research has been conducted on what the introduction of transgenes will mean for centers of domestication. He contends that there is a risk that transgenes will accumulate in native materials, which can lead native plants to serve as "relays" for the unwanted introduction of transgenes to other plants. In addition, Gepts argues, although transgenes are only one among many factors that can reduce genetic diversity, transgenes might cause a severe reduction in genetic diversity. Finally, Gepts notes that we do not know what the environmental effects of genetically modified crops will be in centers of domestication like Mexico, because no one has done the research. Beyond these observations, Gepts considers the relationship of issues related to gene flow, intellectual property, and the culture of agriculture in Mexico, and he offers suggestions for confronting the potential problems posed by gene flow into areas like Mexico.

Kenneth A. Worthy and his colleagues, biologists and social scientists mostly at the University of California at Berkeley, offer a very different perspective on the issues of gene flow. In chapter 8, these authors explore the controversy that erupted after David Quist and Ignacio Chapela published their findings of transgenic material in Mexican maize. Worthy and his colleagues argue that, despite charges that Quist and Chapela's data were questionable as a result of inadequate research technique, their basic finding still stands: There are transgenes in Mexican landraces. Worthy and his colleagues tell us that the debate and review process that surrounded publication of Quist and Chapela's work violated traditional scientific norms of fairness and objectivity. These authors suggest that charges of researcher incompetence, and *Nature*'s public distancing from its original decision to publish Quist and Chapela's paper, may be attributable to "a web of political and financial influence that compromises the appearance of objectivity (and possibly the actual objectivity)" of Quist and Chapela's critics. Worthy and his coauthors trace this web and suggest the dangers to reasonable scientific debate of various forms of conflict of interest.

Dennis Olson's contribution in chapter 9 moves the discussion away from the problem of gene flow and maize. Olson, who is the director of the Trade and Agriculture Project at the Institute for Agriculture and Trade Policy, a Minneapolis-based research, education, and advocacy organization that "promotes resilient family farms, rural communities, and ecosystems around the world," focuses on genetically modified wheat and discusses the problems that this crop poses for U.S. farmers. Although public opposition to genetically modified crops has been muted in the United States, the loud cries of European consumers have led European importers to be cautious about bringing genetically modified crops to the continent, and Olson suggests that the worries of European importers could have a significant influence on the ability of U.S. wheat producers to sell their genetically modified wheat. Olson suggests that European resistance will not affect only farmers who choose to grow genetically modified wheat but other farmers as well, because of the significant possibility that transgenes will accidentally contaminate wheat that has not been genetically modified.

The final contribution to this section comes from Peter Raven, the director of the Missouri Botanical Garden. Raven looks at genetically modified crops in relationship to the serious problem of hunger worldwide. Three billion people around the planet are malnourished, according to Raven, and we must find a way to feed them. Raven contends that genetically modified crops must be part of a multifaceted solution. These crops can improve productivity and the sustainability of agriculture. He says that they can contribute to a reduction in pesticide use and be adapted to radically varying environments. Raven does not deny the process of gene flow but contends that nothing intrinsic to transgenes poses a singular threat. Indeed, according to Raven, current and widely used agricultural practices pose a more serious threat to genetic diversity than does the introduction of genetically modified crops.

Hormone Replacement Therapy and Menopause: Science, Culture, and History

What should be done when a popular prescription drug, used by about six million women in the United States alone, is suddenly found to cause greater health risks than those it was prescribed to treat and prevent? Since the summer of 2002, the millions of women taking hormone replacement therapy (HRT) to treat the symptoms of menopause and reduce the risk of other health problems associated with aging have faced this very question. For decades, doctors have prescribed hormones to menopausal and postmenopausal women, believing that they not only helped with symptoms such as hot flashes but also could protect against cardiovascular disease, osteoporosis, Alzheimer's disease, and even aging of the skin. Pharmaceutical companies promoted HRT as a veritable fountain of youth, capable of preserving beauty and fending off old age. Despite the lack of randomized clinical trials of the drug as a preventative for heart disease, doctors and researchers widely believed that HRT did protect the heart, and physicians often prescribed hormones for long-term use, not just for the short-term treatment of menopause symptoms.

This all changed when the researchers with the Women's Health Initiative (WHI) discovered that, in a randomized clinical trial involving more than sixteen thousand women, the most widely prescribed type of HRT, called Prempro, increased the risk of heart disease, stroke, breast cancer, and dementia. In July 2002, the WHI researchers abruptly halted the study when they found that women taking hormones were slightly more likely than those taking the placebo to have a heart attack. Ten months later, the investigators announced that women taking Prempro, a combination of estrogen and progesterone, were twice as likely to suffer signs of dementia, a finding that flew in the face of a decade of research suggesting that HRT protects the brain and improves cognition. Finally, in 2004, WHI researchers discontinued another segment of the study upon finding that estrogen-only treatments, once thought to carry fewer heart risks than the combined therapy, carried the same increased risk of stroke as the estrogen-progesterone combination.

The case of HRT raises an array of questions about medical research, the pharmaceutical industry, and women's health. For many years, feminist organizations like the National Women's Health Network in the United States raised concerns about HRT, accusing drug companies of promoting hormones before anyone had done sufficient research on their harmful effects. The findings of the WHI seemed to confirm some of these concerns, prompting many to wonder why HRT had been approved and prescribed for so long. Because the study was discontinued, scientists have been discussing the methodological difficulties of studying the benefits and risks of treatments like HRT. To some, the WHI is a "wake-up call," forcing researchers to consider more carefully and humbly the design of drug studies that use observational methods, like those that had earlier indicated the safety and benefits of HRT (Shelton 2002). Dr. Deborah Grady, an expert on HRT at the University of California, San Francisco, has drawn this conclusion from the HRT episode: "A drug should not be used for prevention until it's proven that the benefits exceed the risks in a randomized trial" (quoted in Torassa 2002). On the other hand, many researchers criticize the methods used in the WHI and are continuing to study the role of estrogen in the brain and heart in the hope of finding some benefits (Wickelgren 2003). The results of these studies are not yet known.

For years, critics of HRT have pointed to the role of the pharmaceutical industry in selling the idea of hormone replacement to women and downplaying negative effects. Use of HRT first became widespread after the 1966 publication of *Feminine Forever*, a best-selling book that promoted the myriad benefits of estrogen replacement. The book, which appeared to have been written by an objective doctor, Robert Wilson, was later discovered to have been funded by the Prempro drug maker, Ayerst. Interestingly, the WHI study complicates this view of the relationship between the economic interests of drug makers and the outcome of research. In this case, Wyeth-Ayerst provided funds and free drugs to the WHI and is now facing a dramatic loss of sales as a result. Yet many women still feel, with good reason, that the pharmaceutical industry cannot be trusted because it promoted for years a potentially dangerous treatment as a nearly miraculous disease preventative.

In addition to raising safety concerns about HRT, feminist writers have more generally criticized the "medicalization" of menopause—the treatment of menopause as if it were a disease—and argued instead that it should be understood as natural and normal, not an "estrogen deficiency." The WHI provides some support for this view. Since the WHI was halted, medical researchers on the project have begun to promote alternatives to the long-term use of HRT, recognizing that health problems such as heart disease and dementia cannot be prevented simply by counteracting menopause with hormones. HRT is still considered helpful for alleviating the hot flashes and vaginal dryness associated with menopause in the short term, and the WHI indicates its effectiveness in preventing hip fractures. In regard to colorectal cancer, the data show benefit for women with a uterus who were taking estrogen plus progestin, but no benefit for those who had had a hysterectomy and were taking estrogen alone. WHI researchers have argued that these benefits do not outweigh the risks. The medical professor who led the study, Marcia Stefanick, has stated that it is better to choose alternatives to HRT, suggesting, during an appearance on the *NewsHour with Jim Lehrer* (July 9, 2002), calcium and vitamin D supplementation for preventing hip fractures and colorectal cancer. From this perspective, menopause is no longer treated as though it is a disease, with HRT as a cure-all. Instead, doctors and researchers must now come to new understandings about women's bodies that no longer rely on a "deficiency" model of aging.

The authors in this volume bring together a diverse set of perspectives on HRT and on drug testing more generally. In chapter 11, Dr. Sylvia Wassertheil-Smoller, chief of the Women's Health Initiative at the Albert Einstein College of Medicine, introduces the section on HRT with a thoughtful explanation of the WHI, its methods, findings, and critics. Emphasizing that menopause is not a disease, Wassertheil-Smoller argues that the WHI provided, for the first time in decades, sufficient information about the risks and benefits of HRT so that women can make informed choices about how to lessen menopausal symptoms and prevent diseases associated with aging.

In chapter 12, Judith Houck presents a historian's perspective of medical approaches to menopause, tracing how medicine changed the experience and meaning of menopause from the late nineteenth century to the present. Tracing several phases in the medical understanding of menopause, Houck identifies the current phase as one of the medicalization not simply of menopause but of female aging in general. This resonates with current critiques of the prescription of HRT for long-term use as a preventative of a range of diseases associated with aging. Yet Houck takes issue with critiques of medicalization that do not acknowledge that the majority of American women older than fifty have not used HRT and that many have sought alternative treatments and approaches to menopause. Although drug companies and physicians have for decades encouraged medical treatments for menopausal women, many women have not embraced this view of their changing bodies as "deficient" and instead have creatively sought ways of dealing with menopause.

In chapter 13, Barbara Seaman, an author well known for her work in the women's health movement and numerous books on the birth control pill and sex hormones, sketches a timeline of events in the development of HRT, emphasizing the role of the pharmaceutical industry in promoting the use of these drugs, despite research that indicated negative health effects. She argues that the harmful effects of hormone therapies were documented long before the Women's Health Initiative studies, yet these dangers were sidelined as a result of deceptive industry campaigns in favor of HRT and enthusiasm for the drugs in the popular media. Her study is alarming, forcing us once again to wonder why women have been exposed to these risks for so many years and whether the medical industry is exposing women to similar unrecognized risks today.

Margaret Lock presents a fascinating comparative view of how women experience menopause in different regions of the world, arguing in chapter 14 that biology and culture are interdependent in complex ways. Of course, all women experience the cessation of menstruation, yet the symptoms of this change, and the health problems often associated with it, are not the same around the world. For example, while women in North America and Europe most commonly report hot flashes and night sweats, Japanese women are much more likely to report shoulder stiffness. Lock draws on a concept of "local biologies" to refer to the way that the physical experience of menopause is "contingent on evolutionary, historical, cultural, and individual variables." Women draw upon local cultural categories of knowledge in their experiences of menopause, producing enormous variation and complexity in the experience of what is often assumed to be a universal biological fact.

Readers interested in the issues of research design and methodology raised in Wassertheil-Smoller's chapter in this section should turn to chapter 15, by David L. DeMets. DeMets, an expert in medical research methodology, discusses the connections between clinical studies like the WHI and the realities of medical practice. Using HRT as a case study, DeMets demonstrates the importance of clinical trials, as opposed to observational studies, in evaluating the benefits and consequences of a new drug. Although clinical trials may be costly and time consuming, without these rigorous tests patients may be exposed to unknown risks for unsubstantiated reasons.

Smallpox and Bioterrorism

Naturally occurring smallpox, a highly contagious and often fatal disease, was eradicated in 1977, through an intensive campaign of immunization around the world. So why is smallpox a public health concern again today? During the Cold War, the United States and the Soviet Union retained the smallpox virus in storage. Because these samples of the virus continue to exist, there is a possibility that the smallpox virus, or a modified strain of it, could be used as a biological weapon. In the context of heightened concerns about terrorism, the Bush administration called for dramatic steps to prevent the devastating effects of such an attack. The

federal government aimed to vaccinate some 450,000 people, but ultimately only about thirty-eight thousand people were vaccinated nation-
wide (Meckler 2003).

Scientists and others raised questions about the real risks of smallpox,
particularly in comparison to existing epidemics that demand public
health resources. Some scientists have argued that smallpox is not as contagious and difficult to control as it has been portrayed (Enserink 2002).
Others note that the vaccine itself poses serious risks. Donald Hopkins, a
world-renowned expert on global public health, says that the main risk of
vaccinating the public against a potential smallpox epidemic is that "some
of those vaccinated will have severe reactions to the vaccine, including
some deaths," and expresses concern that by "reintroducing smallpox
again in the form of vaccinations . . . we are losing a small part of the benefit of having eradicated smallpox in the first place" (Carter Center 2002).
In light of concerns such as these, the smallpox vaccination campaign has
largely fizzled out, as David Rosner and Gerald Markowitz describe in
chapter 18. Many scientists agree that a better approach is to provide vaccinations only during an outbreak, because, as Dixie Whitt explains in
chapter 16, the incubation period of smallpox is long enough that someone immunized within two to three days of exposure should develop an
adequate immune response. This approach is now endorsed by the U.S.
Centers for Disease Control and Prevention.

For such a quick response to a bioterrorist attack—or to any new epidemic—our public health institutions must be well prepared. Recent
public health concerns, such as severe acute respiratory syndrome (SARS)
and West Nile virus, in conjunction with the realization that bioterrorism
is possible, whether with anthrax or some other disease, have prompted
broad debate about the state of public health institutions in the United
States. Four chapters take on the complexities of dealing with the public
health risk of an intentional smallpox outbreak or other bioterrorist attack.

As two of our authors note, smallpox would not be an issue today if the
United States and Soviet Union had not preserved samples of the virus
during the Cold War. It would be a mistake to discuss the threat of smallpox and other forms of bioterrorism without considering why biological
weapons remain a threat today. Some have pointed a finger at the U.S.
government for resisting agreement on an inspection regime designed to
strengthen the Biological Weapon Convention (Wright 2002; Rosenberg
2002), and for engaging in research that some say comes too close to pro-

ducing biological agents that could be used offensively (FAS Working Group 2003).

Without tougher international laws and a commitment to inspections, it seems that the threat of biological weapons will remain with us. This leaves our public health institutions as a first line of defense. Now is a crucial time to build our understanding of potential (and existing) epidemics and to debate the role of public health institutions in preventing and responding to these new diseases.

Dixie Whitt, a microbiologist, introduces the section on smallpox, bioterrorism, and public health with a discussion of how the smallpox disease and vaccine work in the human body. By placing our knowledge of smallpox in historical context, Whitt effectively explains why we do not know a lot about smallpox and argues that much of today's controversy surrounding smallpox stems from these gaps in knowledge. By combining an up-to-date summary of the state of knowledge with a fascinating history of its development, Whitt provides the background necessary for discussing the complex and often controversial debates about how to protect ourselves from the threat of smallpox bioterrorism.

After the terrorist attacks on September 11, 2001, and the subsequent anthrax scare, a group of scholars at the Center for Law and the Public's Health at Georgetown and Johns Hopkins Universities drafted a model law intended to help the states deal with public health emergencies, such as a potential bioterrorist attack. This model law spurred a great deal of debate among public health scholars and practitioners, lawmakers, civil liberties activists, and a diverse range of interest groups, including AIDS activists, who worried that the proposed quarantine rules could be used to target those with compromised immune systems. In chapter 17, Lesley Stone, Lawrence Gostin, and James Hodge, who helped draft the model legislation, explain the purpose of the draft law and counter the arguments of its diverse critics. Arguing that existing state public health laws would have been insufficient to cope with a bioterrorist attack or other extreme public health emergency, Stone, Gostin, and Hodge suggest that adoption of the model legislation is crucial to both protecting the public's health and safeguarding individual rights.

In chapter 18, David Rosner and Gerald Markowitz provide the perspective of public health officials. Focusing on two crucial events—the smallpox inoculation campaign and the circulation of the draft Model State Emergency Health Powers Act—Rosner and Markowitz argue that, despite early hopes of reforming the nation's public health services, these crucial programs for bioterrorism preparedness are now endangered. For one, public health officials are finding that the emphasis on smallpox detracted from other important public health issues. Second, discussions of the draft legislation for state public health law highlighted concerns about the growing power of the state in the post–September 11 environment. Combined with state budget crises, differences in perception of public health needs at the state and federal levels have undermined popular public health programs.

In chapter 19, Judith Walzer Leavitt provides perspective on current strategies for public health emergencies with a historical analysis of two outbreaks of smallpox in major American cities, Milwaukee and New York. Leavitt identifies a number of characteristics that can make the difference between a successful public health response and a devastating failure. Most crucially, Leavitt finds that coercion, whether for quarantine or vaccination, has led not to the reduction of the smallpox threat but rather has increased the problems and public resistance. Based on the historical evidence, she advocates approaches that are culturally sensitive and cooperative, rather than forced from the top down.

Conclusions

The four sections of this book are intended to be separate, with disparate issues spanning a range of science and technology. But some themes woven through all are notable. They all illustrate the challenge and accompanying risks of making decisions in the absence of sufficient information. Although everyone agrees that policy should be evidence-based, different constituents advocate making decisions at different points on the continuum between ignorance and conclusive knowledge. Another theme illustrated by all the sections is that science and technology exist in a global context. Antibiotic use in the United States affects decisions in Australia, closure of European markets to genetically engineered products affects U.S. farmers, and storage of smallpox virus in the United States and Russia influences bioterrorism and responses to it worldwide. Finally, all the chapters show us how deeply each of us is affected by the challenges and opportunities of science and technology. Each of us is required to make frequent personal choices about our foods and drugs. And, not infrequently, we are confronted with dramatic outcomes of our personal choices and those of others—a death from antibiotic-resistant bacteria, a stroke that might have been caused by hormone replacement therapy, an evacuation caused by a threat of bioterrorism, or a farm foreclosure caused by an international response to genetically engineered crops. Universally, these issues touch us as members of a pluralist society in which we share responsibility for educating ourselves and each other so that we can maintain vigorous debate about issues of science and technology in our society.

References

Action Group on Erosion, Technology and Concentration. 2002. Ag biotech countdown: Vital statistics and GM crops. *Geno-Types* (June).

Austin, D. J., K. G. Kristinsson et al. 1999. The relationship between the volume of antimicrobial consumption in human communities and the frequency of resistance. *Proceedings of the National Academy of Science USA* 96(3): 1152–56.

Carter Center. 2002. The greatest killer: Smallpox in history: A Q&A session with Carter Center's Dr. Donald Hopkins on the eradication of smallpox. August 22. http://www.cartercenter.org/viewdoc.asp?docID=1045&submenu=news. (accessed August 3, 2004).

Conway, S. P., K. G. Brownlee, et al. 2003. Antibiotic treatment of multidrug-resistant organisms in cystic fibrosis. *American Journal of Respiratory Medicine* 2(4): 321–32.

Clark, N. M., E. Hershberger, et al. 2003. Antimicrobial resistance among gram-positive organisms in the intensive care unit. *Current Opinions in Critical Care* 9(5): 403–12.

Enright, M. C. 2003. The evolution of a resistant pathogen—The case of MRSA. *Current Opinions in Pharmacology* 3(5): 474–79.

Enserink, Martin. 2002. How devastating would a smallpox attack really be? *Science* 296 (May 31): 1592–95.

FAS Working Group on Biological Weapons. 2003. Position paper: Secret biodefense activities are undermining the norm against biological weapons. January 2003. http://64.177.207.201/cbw/papers/index.htm (accessed August 3, 2004).

Gaskell, George et al. 2001. Troubled waters: The atlantic divide on biotechnology policy, pp. 96–115. In George Gaskell and Marin W. Bauer, eds., *Biotechnology, 1996–2001: Years of controversy.* London: Science Museum.

Greenpeace. 2004. GE agriculture and genetic polution. http://www.greenpeace.org/international_en/campaigns/intro?campaign_id=3997 (accessed Oct. 1, 2004).

Hiramatsu, K., K. Okuma et al. 2002. New trends in *Staphylococcus aureus* infections: Glycopeptide resistance in hospital and methicillin resistance in the community. *Current Opinions in Infectious Diseases* 15(4): 407–13.

Howard, Ted and Jeremy Rifkin. 1977. *Who should play God? The artificial creation of life and what it means for the human race.* New York: Delacorte.

Interagency Task Force on Antimicrobial Resistance. 1996. Antibiotic use in hospitals. A public health action plan to combat antimicrobial resistance, OTA.

Kloppenburg, Jack R. 2004. *First the seed: The political economy of plant biotechnology, 1492–2000.* 2d ed. Madison: University of Wisconsin Press.

Lederberg, J. 2000. Infectious history. *Science* 288(5464): 287–93.

Mackenbach, J. P. and C. W. Looman. 1988. Secular trends of infectious disease mortality in the Netherlands, 1911–1978: Quantitative estimates of changes coinciding with the introduction of antibiotics. *International Journal of Epidemiology* 17(3): 618–24.

Meckler, Laura. 2003. Program to vaccinate health workers against smallpox is faltering, officials say. *Associated Press State & Local Wire.* July 24. http://web.lexis-nexis.com/universe.

Mehta, R. M. and M. S. Niederman. 2003. Nosocomial pneumonia in the intensive care unit: Controversies and dilemmas. *Journal of Intensive Care Medicine* 18(4): 175–88.

Price, D. B., D. Honeybourne et al. 2004. Community-acquired pneumonia mortality: A potential link to antibiotic prescribing trends in general practice. *Respiratory Medicine* 98(1): 17–24.

Projan, S. J. 2003. Why is big pharma getting out of antibacterial drug discovery? *Current Opinions in Microbiology* 6: 427–30.

Rosenberg, Barbara Hatch. 2002. Anthrax attacks pushed open an ominous door. *Los Angeles Times.* September 22. http://www.anthraxinvestigation.com/anthraxreport.htm (accessed August 3, 2004).

Shelton, James D. 2002. Humility in observational studies (in letters). *Science* 297 (September 27): 2208.

Torassa, Ulysses. 2002. Warnings from the scientific community: Estrogen-progestin risk took decades to uncover; small danger to individual women hid larger pattern. *San Francisco Chronicle,* July 14, p. A3.

Weigel, L. M., D. B. Clewell et al. 2003. Genetic analysis of a high-level vancomycin-resistant isolate of *Staphylococcus aureus. Science* 302 (5650): 1569–71.

Wickelgren, Ingred. 2003. Brain researchers try to salvage estrogen. *Science* 302 (November 14): 1138–39.

Wilson, J. F. 2002. Renewing the fight against bacteria. *cientist* 16(5):22.

Wright, Susan, ed. 2002. *Biological warfare and disarmament: New problems/new perspectives.* Lanham, Md.: Rowan and Littlefield.

PART 1
Overuse of Antibiotics on the Farm

1

Antibiotic Resistance

The Agricultural Connection

Christine Mlot

Dairy cows with symptoms of mastitis typically receive penicillin to stop the infection. Similarly, farmers administer tetracycline to swine and poultry to suppress respiratory diseases. Bacterial infections in catfish on fish farms and in pear trees in orchards are treated with tetracyclines or other antibiotics too. And to boost their rates of growth, healthy livestock routinely consume small doses of antibiotics in manufactured animal feeds.

As the Stanford University microbiologist Stanley Falkow has remarked, the biosphere is, in effect, bathed in a dilute solution of tetracycline, and roughly half of the one million metric tons of antibiotics (Davies 2000) in the solution have entered through the food supply (World Health Organization 2000). Most conventionally grown poultry, pork, and beef in the United States is produced with the help of antibiotics at some point. As with humans, animals are given the drugs to eliminate infections or prevent imminent ones. More controversial is the routine use of small but long-term doses of antibiotics to hasten the animals' growth to market size. Such use accounts for most of U.S. agriculture's share of antibiotics (Levy 1998; Mellon, Benbrook, and Benbrook 2001).

The use of antibiotics in agriculture, producer groups say, has given consumers an abundant and affordable supply of food, especially meat. But the intensive use of antibiotics comes with a cost. Antibiotics kill bacteria or stop them from growing by disrupting the cells' machinery. At the same time, antibiotic use selects for bacterial cells containing genes that allow the cells to disable or evade the drugs, most of which are also used to treat human infections. The resistance genes can be transferred through foodborne bacteria or other means to the human intestine and have been linked to drug-resistant infections in people and even to several deaths (World Health Organization 2000). Although the misuse of anti-biotics in human medicine is also a big contributor to the development of resistance—an estimated half of antibiotics prescribed are unnecessary (Levy 1998)—agriculture's contribution cannot be dismissed.

Public health–monitoring programs are turning up increased fre-quencies and spread of resistant bacteria that cause foodborne illness, particularly *Salmonella, Campylobacter,* and toxic strains of the normally benign *Escherichia coli* (U.S. General Accounting Office 1999). Agricul-tural use of one class of antibiotic has also been linked to development of resistance to the human version of the drug among enterococci, a group of potentially deadly hospital-acquired pathogens (Kaufman 2000).

In response to these trends, researchers are looking for ways to reduce the use of antibiotics in agriculture, particularly their use as growth pro-moters. Remedies range from renewed attention to sanitary animal hus-bandry and hygiene to experimental immunological methods and truly novel forms of antimicrobials. But progress in moving alternatives to the farm has been slow and, as a recent federal plan to combat antibiotic re-sistance notes, requires an intensified research and regulatory push (Task Force 2001).

Ecology and Policy

Bacteriologists since Alexander Fleming, who discovered penicillin in 1928, have worried about the selective force of antibiotics and the result-ing problems of resistance. Particularly on the agricultural front, however, the development of public policy to minimize the development of resist-ance has been lumbering. As early as 1978, the U.S. Food and Drug Ad-ministration (FDA) attempted to eliminate the agricultural use of some antibiotics that are also used in human medicine (U.S. General Account-ing Office 1999). But producer and pharmaceutical groups resisted new regulations, arguing that no definitive evidence existed of harm to humans from antibiotic use in animals. The weight of evidence linking the two has since steadily increased, along with a better understanding of the nature of resistance.

Figure 1.1 One relatively little known use of antibiotics is on fruit trees, for the prevention of such diseases as fireblight. Although plant use accounts for only 0.1 percent of total antibiotic use, researchers are developing alternatives. These pear cultivars have been bred to withstand infection by the *Erwinia* bacteria that cause fireblight. Courtesy Agricultural Research Service, USDA.

In 1998 the FDA's Center for Veterinary Medicine responded with a proposal for evaluating and curtailing unnecessary drug use in animals to curb resistance (FDA 1999). Critics pointed out that the proposal did not address existing uses of antibiotics and that the agency has been slow to implement the proposal (Task Force 2001). But the proposal broke new ground for the agency in acknowledging the connection to human health. "FDA is convinced that there is more than adequate scientific evidence demonstrating that resistance develops in enteric pathogens in animals in response to drug use and that they can be transferred through the food supply to humans," said the FDA's Linda Tollefson at the May 2000 meeting of the American Society for Microbiology.

The FDA—along with the Centers for Disease Control and Prevention, the National Institutes of Health, and other agencies, including the U.S. Department of Agriculture (USDA)—also signed on to a plan, first issued in the summer of 2000, that attempted to coordinate the federal response to antimicrobial resistance (Task Force 2001). The task force grew out of 1999 Senate hearings, cochaired by Bill Frist, R-Tenn., and Edward Kennedy, D-Mass., on the wide scope of antibiotic resistance and, consequently, the multiple federal agencies and departments that need to address the issue. The report by the Task Force on Antimicrobial Resistance spelled out eleven steps for action, including a call to monitor drug use in agriculture as well as in human medicine and consumer goods. It also nudged the FDA to put into place its proposed system for approving drug use in agriculture and to reevaluate drugs currently in use.

Congress has taken further note of the issue of antibiotic resistance arising from agricultural use. Members of both the House and Senate introduced a bill, called the Preservation of Antibiotics for Human Treatment Act of 2002, which would phase out the use of certain antibiotics as growth promoters in animal feed.

Yet even the most stringent national policy is not enough, given the ease with which both people and microbes move around the globe. The interagency report aims to address international concerns, noting that the United States already lags behind other developed countries in regulating agricultural use of antibiotics. As the World Health Organization has recommended, the European Union has eliminated the controversial use of the same antibiotics as both human medicine and growth promoters for animals, with little adverse effect on the economy or public health.

Denmark has voluntarily given up use of all antibiotics as growth promoters. Among developed countries, only the United States and Canada still allow chlortetracycline, oxytetracycline, and penicillin, crucial for treating humans, to also be used on animals (Angulo 2000b).

Developing countries have even fewer guidelines for proper use of antibiotics, at the same time that many are increasing livestock production. They are also battling many other bacterial diseases rarely seen in developed countries, such as food- and waterborne *Shigella dysenteriae*, which is now resistant to most drugs available (World Health Organization 2000). The World Health Organization (2000) points out how globalization of trade and travel all but ensures the spread of resistant microbes. A single clone of *Salmonella* known as DT104 emerged in England in the 1980s and is now considered to be at pandemic levels. In its hegemonic spread, the pathogen has acquired resistance to at least five kinds of antibiotics.

Paths of Resistance

Collectively, bacteria have evolved the means to dispose of each of the more than 150 antibiotics (World Health Organization 2000) developed by modern medicine. Bacterial resistance genes code for beta-lactamases, transferases, and other enzymes that have been altered through the course of bacterial evolution and now interfere with the drugs' actions. The enzymes may break down the drug or alter its target in the bacterial cell. The bacteria may also simply eject the antibiotic.

Resistance genes can be passed around inside an animal's intestine, within closely related species or strains of bacteria, and across genera through horizontal gene transfer. The genes are commonly carried on extrachromosomal plasmids that get transferred to other microbes through conjugation. When transposons, or jumping genes, bundle other resistance genes into a plasmid, the receiving microbe becomes equipped with multiple resistances. Viruses, or bacteriophages, can also infect other bacteria with resistance genes, at least in the lab (Salyers 2000).

"The spread of genes is the problem, not just the spread of bacteria," according to the microbiologist Abigail Salyers (n.d,; see also chapter in this volume). Resistance genes spawned on the farm can circulate via several avenues. In a 1976 paper in *Nature*, Stuart Levy, G. B. FitzGerald,

Figure 1.2 Large-scale poultry producers raise chickens in flocks of a million or more. The birds can spend 42 of their 45-day lives on antibiotics. Courtesy Agricultural Research Service, USDA.

and A. B. Macone documented the movement, among chickens caged together, of a gene for antibiotic resistance. They also isolated *E. coli* containing the gene from two farmworkers, indicating the transfer of bacteria from animals to humans whether by air or direct handling of the animals.

Antibiotic-resistant *Salmonella* that infected a twelve-year-old farm boy probably followed the same path (Fey et al. 2000). Writing in the *New England Journal of Medicine,* researchers detailed how they used DNA and enzyme analysis to show that the strain of *Salmonella enterica* serotype typhimurium was identical to bacteria taken from a cow on the farm that had been treated for *Salmonella* a month before the boy became ill. Tests showed that the bacteria were resistant to thirteen drugs, including ceftriaxone, an "expanded-spectrum" cephalosporin used for humans (Fey et al. 2000). Ceftriaxone resistance had been reported in other countries but not in the United States (Fey et al. 2000). Although ceftriaxone is not approved for use on cattle, other expanded-spectrum cephalosporins are, leading the researchers to conclude that antibiotic use on livestock is primarily responsible for generating resistance in *Salmonella.*

Apart from the direct path between farm animals and farmworkers, antibiotic-resistant bacteria can contaminate food during slaughter and make their way to the consumer's plate and intestines through under-cooked food. Public health officials have traced this path by following the recent increase in fluoroquinolone resistance among *Campylobacter* in chickens and people (Angulo 2000b).

Although most people can weather a bout of *Salmonella* or *Campylobacter* food poisoning, infections can turn serious if the bacteria invade the blood. More than twenty-five thousand people in the United States are hospitalized with such serious foodborne infections, and hundreds die from them each year (U.S. General Accounting Office 1999; Angulo 2000a). Widely introduced for human use in 1988, the costly fluoroquinolones have been a key treatment for serious *Salmonella* and *Campylobacter* infections. Despite protest from public health officials (Angulo 2000a), in 1995 the drugs were approved for treating respiratory disease in chickens in the United States.

Researchers predicted that fluoroquinolone resistance would increase, and it did (Angulo 2000a). Although *Campylobacter* is a pathogen in humans, it is a benign microbe in chickens. It turns up in 60 to 80 percent of chickens packaged for sale in U.S. markets, according to a study by the Centers for Disease Control and Prevention. In 1998, 20 percent of the chickens on supermarket shelves carried fluoroquinolone-resistant strains, and 13 percent of human *Campylobacter* infections were resistant to the drugs. By contrast, a 1991 study of isolates from human *Campylobacter* infections turned up no fluoroquinolone resistance. In 1999, the proportion of resistant human infections rose to 20 percent (Angulo 2000b).

The increase in drug-resistant human infections, along with increasing public health concern and media attention, prompted the FDA to reverse its 1995 approval for certain agricultural use of the drugs. In 2000 the agency began the process of withdrawing approval of fluoroquinolone use for poultry (Food and Drug Administration 2000).

Fluoroquinolone resistance has also appeared in *Salmonella* (Pittock 2002). About 95 percent of the *Salmonella* DT104 strains are resistant to ampicillin, chloramphenicol, streptomycin, sulfonamides, and tetracycline (Angulo 2000b), and resistance to one kind of antibiotic often begets resistance to others. Resistance monitoring programs have recently detected isolates of a second *Salmonella* strain that resists at least eight classes of antibiotics (Gupta et al. 2003).

Some enterococci have also acquired resistance to all antibiotics, including vancomycin, considered an important antibiotic of last resort. Widespread hospital use of vancomycin has selected for vancomycin resistance in these bacteria (VRE), which can be deadly for immuno-compromised patients. But epidemiological studies have also connected agricultural antibiotic use to the rise of VRE in some areas (U.S. General Accounting Office 1999).

Vancomycin is a glycopeptide, as is the veterinary drug avoparcin, and resistance to one will confer resistance to the other. Although avoparcin has not been allowed for animal use in the United States, until 1997 it was used as a growth promoter for livestock in Europe. On farms where avoparcin was used but not on other farms, researchers were able to isolate VRE from poultry and pig feces. In parts of Germany where vancomycin is rarely used in hospitals, VRE have been found in grocery meats and in people in the community at large. In contrast, because avoparcin is not used in the United States, VRE are rarely found outside hospitals. And after use of avoparcin was banned as a growth promoter, studies from Denmark and Germany indicate, the number of VRE isolates declined in both poultry and the community (World Health Organization 2000).

To combat the resistance of enterococci to vancomycin, in the fall of 1999 the FDA approved a new last-resort antibiotic. Synercid became the first human antibiotic in the class of streptogramins, but some bacteria were already primed to withstand it. An animal version of the drug, called virginiamycin, has been used in chickens and other animals to treat disease and as a growth promoter. And bacteria with a gene that confers resistance to macrolides, another class used in human medicine, have cross-resistance with streptogramins. All these uses, human and animal, seem to have contributed to a pool of genes resistant to the new drug. Looking at chickens in which virginiamycin has been used, researchers have found significant numbers of enterococci that show resistance to Synercid (Zerno 2000). Low levels of Synercid resistance have also turned up in people, including a woman who died at a Michigan hospital after the drug failed to clear her VRE infection (Kaufman 2000).

Researchers are concerned that other reservoirs of bacteria harbor resistance genes that can get passed on to human pathogens. Resistance genes have turned up in everything from bacterial cultures of cheese—the resistance genes probably originated with antibiotics given to cows—to the skin flora of people who have been treated with antibiotics for acne

(Eady 2000). And the facile movement of genes in the microbial world is not just a one-way path. Resistance genes arising from the overuse of antibiotics in humans may in turn be affecting agriculture. The use of treated sewage as fertilizer on fields, for example, may be a vector for spreading to animals resistance spawned in humans.

Alternatives

When an antibiotic is not in use, bacteria without resistance genes probably will be more abundant than bacteria with resistance genes, because the enzymes and other proteins used to dismantle the antibiotic cost energy to produce. Therefore nonresistant bacteria will have an advantage. "Susceptible bacteria should be our teammates in confronting and reversing the resistance problem," Levy has written (1997, 7). Vancomycin-resistant bacteria in livestock decreased by half within two years after the European Union banned avoparcin. Similarly, tetracycline-resistant *E. coli* in human infections dropped to half their previous levels in East Germany five years after the East Germans banned the use of tetracycline animal feeds in the 1980s (cited in Witte 1997). Still, although resistance levels may drop with the removal of an antibiotic, they never disappear altogether and would likely rebound if use of the antibiotic were reinstated.

Consequently, researchers are looking for alternatives to the conventional uses of antibiotics, both in controlling agricultural pathogens and in boosting livestock growth. The possibilities range from high tech to low. The reports of the Task Force on Antimicrobial Resistance (2001) and the World Health Organization (2000) emphasize sound animal husbandry—controlling infections by separating sick animals and their feed and water, washing implements and hands, and taking other hygienic steps. "In the absence of successful infection control, no amount of prudent antimicrobial use will be successful in avoiding further dissemination and resistance development," says Dale Hancock (2000), a professor of veterinary medicine.

Bacteria themselves are being enlisted to combat pathogen spread and antibiotic resistance. Normal flora are a natural part of an animal's defenses, and certain strains are especially effective in warding off pathogens (see, for example, Casas and Dobrogosz 2000). Animal producers have begun to use specially tested cultures of bacteria, called probiotics, that can

Figure 1.3 Separating sick animals and their food and water from healthy ones can curb the spread of pathogens and reduce the use of antibiotics in livestock. Pictured here are Holstein dairy cows. Courtesy Agricultural Research Service, USDA.

prevent pathogens from colonizing. Some cultures come from disease-free animals and are undefined, whereas others are formulated according to a specific recipe. The FDA has approved one product for chickens that prevents contamination by *Salmonella* and other pathogens (USDA 1998). The mix of twenty-nine beneficial bacteria colonizes newly hatched chicks, excluding potentially pathogenic types. Other cultures are being devised to compete with plant pathogens and reduce the need for spraying fruit trees with antibiotics (McManus 2000).

Surprisingly, some probiotics might also work as growth promoters (Casas and Dobrogosz 2000). Exactly how antibiotics promote animal growth is still not understood, despite fifty years of use. Some studies suggest that low doses of antibiotics suppress intestinal bacteria that siphon nutrients from the animal, because germ-free animals don't benefit from antibiotics used for growth promotion (Curtiss 2000). Understanding how growth promoters work should reveal other ways to achieve the same effect, maintains Roy Curtiss III (2000), a microbial geneticist.

Tweaking the immune system may provide ways to stimulate growth as well as to stave off infections. One method is to create antibodies against targets that regulate growth. Researchers have experimentally enhanced the growth rate of pigs by using antibodies that target a peptide component of a growth hormone (Yancey 2000). In chickens, using antibodies to block the effect of cytokines, which can cause wasting, results in faster animal growth (Cook 2000). Chicken interferon gamma is being field tested as an alternative to antibiotic therapy for infections and for weight gain (Lowenthal et al. 2000). In a similar study, researchers are looking at passive antibody therapy as a way to prevent infections in piglets (Normantiene et al. 2000).

Several livestock vaccines designed to prevent infections are on the market and others are in the works. An attenuated *Salmonella* vaccine for chickens reduces colonization by the pathogen. Another is in development for cattle. Some experimental vaccines have been developed for mastitis, which affects half of all dairy cows each year, but getting vaccines to work on a herd level is difficult (Yancey 2000).

Researchers also hope to exploit another arm of animals' natural defenses. Animals across the kingdom produce small antimicrobial peptides; these are known as defensins in humans, for example, and magainins in frogs. Such natural microcidal molecules often work by physically disrupting bacterial cell membranes, a mechanism that may thwart development of resistance (Zhang, Ross, and Blecha 2000). Pigs, for example, produce more than a dozen antimicrobial peptides, which are being studied for prospective drug development (Zhang, Ross, and Blecha 2000).

In addition to searching for new antibiotics, particularly truly novel ones, researchers are looking at ways to keep the available ones effective. One new strategy is to prevent bacteria from pumping out antibiotics. Tufts University researchers are developing a compound that prevents bacteria from expelling tetracycline, reviving the drug's effectiveness (Levy 1998). Another new drug on the market inhibits the bacterial enzyme that breaks down penicillin (Levy 1998). Physicians and veterinarians alike are also discovering that shortened courses of antibiotics may be effective in clearing certain bacterial infections while reducing the selective pressure for resistance that comes with traditional long-term courses (Hancock 2000; Albanese 2000).

The recent federal plan to combat antimicrobial resistance also highlights further development of bacteriophages as a promising therapy to treat foodborne pathogens (Task Force 2001). The Task Force on Antimicrobial Resistance recommends irradiating food to reduce the spread of foodborne pathogens, although this may require overcoming consumer resistance.

Most pressing for public health officials, including those at the World Health Organization, is avoiding the use of, or finding alternatives to, the use of antibiotics from the human medicine cabinet as growth promoters. One recommendation is for greater use of the class of antimicrobials known as ionophores, which have no human medicine analog and can serve as growth promoters (Crawford 2000).

Economics probably will dictate which, if any, alternatives are adopted. "Food animal production is a business," and a competitive one, notes Robert Yancey Jr. of the drug company Pfizer (2000). But the European model and consumer demand may be spurring agricultural producers to adopt alternatives to antibiotics. Newly formulated USDA rules for organic farms expressly prohibit the use of antibiotics as growth promoters in livestock (USDA 2002). While organic farming so far accounts for a tiny part of the total food acreage in the United States, it is booming, having doubled in acreage during the 1990s (Dimitri and Greene 2002).

Bigger market forces also are at work to reduce the use of antibiotics in agriculture. In 2003, the fast food giant McDonald's announced that it would stop its meat producers from using antibiotics as growth promoters for animals (Kaufman 2003). This follows similar policies by other fast food chains. The chains' cumulative and global reach may very well translate into widespread use of some alternatives to heavy agricultural use of antibiotics.

At the same time the ability of bacteria to evolve means that the search for alternatives will no doubt continue.

Note

An earlier version of this piece appeared in *BioScience* 50 (November 2000): 955–60.

References

Albanese, Bernadette. 2000. Johns Hopkins University. Personal communication. August 10.

Angulo, Fred. 2000a. Centers for Disease Control and Prevention. Personal communication. August 17.

———. 2000b. Presentation at annual meeting of the American Association for the Advancement of Science. Washington, D.C., February 19.

Casas, Ivan A., and Walter J. Dobrogosz. 2000. Validation of the probiotic concept: *Lactobacillus reuteri* confers broad-spectrum protection against disease in humans and animals. *Microbial Ecology in Health and Disease* 12:247–85.

Cook, Mark E. 2000. The interplay between modern management practices and the chicken: How immune response and the physiological mechanisms for growth and efficiency have adapted over time. Where do we go from here? Pp. 97–110. In *Biotechnology in the Feed Industry,* edited by T. P. Lyons and K. A. Jacques. Nottingham, United Kingdom: Nottingham University Press.

Crawford, Lester. 2000. Georgetown University. Personal communication. March 16.

Curtiss, Roy III. 2000. Presentation at 100th general meeting of American Society for Microbiology. Los Angeles. May 24.

Davies, Julian. 2000. Presentation at annual meeting of the American Association for the Advancement of Science. Washington, D.C. February 20.

Dimitri, Carolyn, and Catherine Greene. 2002. Recent growth patterns in the U.S. organic foods market. ERS Agriculture Information Bulletin No. AIB777. http://www.ers.usda.gov/publications/aib777/aib777a.pdf (accessed March 17, 2004).

Eady, A. 2000. Presentation at 100th general meeting of American Society for Microbiology. Los Angeles. May 24.

Falkow, Stanley. 2000. Stanford University. Personal communication. August 9.

Fey, Paul D. et al. 2000. Ceftriaxone-resistant *Salmonella* infection acquired by a child from cattle. *New England Journal of Medicine* 342(17): 1242–49.

Gupta, A. et al. 2003. Emergence of multidrug-resistant *Salmonella enterica* serotype Newport infections resistant to expanded-spectrum Cephalosporins in the United States. *Journal of Infectious Diseases* 188:1707–16.

Hancock, Dale. 2000. Presentation at 100th general meeting of the American Society for Microbiology. Los Angeles. May 22.

Kaufman, Marc. 2000. Worries rise over effect of antibiotics in animal feed. *Washington Post,* March 17.

———. 2003. McDonald's will tell meat suppliers to cut antibiotics use. *Washington Post,* June 18.

Levy, Stuart B. 1997. Antibiotic resistance: An ecological imbalance. pp 1–9 in *Antibiotic resistance: origins, evolution, selection, and spread,* edited by Derek J. Chadwick and Jamie Goode. *Ciba Foundation Symposium* 207. New York: John Wiley and Sons.

———. 1998. The challenge of antibiotic resistance. *Scientific American* (March): 46–53.

Levy, Stuart B., G. B. FitzGerald, and A. B. Macone. 1976. Spread of antibiotic resistance plasmids from chicken to chicken and from chicken to man. *Nature* 260:40–42.

Lowenthal, J. W. et al. 2000. Avian Cytokines: The natural approach to therapeutics. *Development & Comparative Immunology* 24(2–3): 355–65.

McManus, Patricia S. 2000. Presentation at annual meeting of the American Association for the Advancement of Science. Washington, D.C. February 19.

Mellon, Margaret, Charles Benbrook, and Karen Lutz Benbrook. 2001. *Hogging it! Estimates of antibiotic abuse in livestock.* Cambridge, Mass.: Union of Concerned Scientists Publications. http://www.ucsusa.org/food_and_environment/antibiotic_resistance/page.cfm?pageID=264 (accessed March 13, 2004).

Normantiene, T., V. Zukaite, and G. A. Biziulevicius. 2000. Passive antibody therapy revisited in light of the increasing antibiotic resistance: Serum prepared within a farm reduces mortality of dystrophic neonate piglets. *La Revue de Médecine Vétérinaire* 151 (2): 105–8.

Pittock, Laura J. V. 2002. Fluoroquinolone resistance in *Salmonella* serovars isolated from human and food animals. *FEMS Microbiology Reviews* 26:3–16.

Salyers, Abigail. n.d. Agricultural use of antibiotics and antibiotic resistance in human pathogens: Is there a link? http://web.archive.org/web/20010222073241/http://www.healthsci.tufts.edu/apua/salyerschapter.htm (accessed April 21, 2000).

———. 2000. Personal communication. August 15.

Task Force on Antimicrobial Resistance. 2001. *A public health action plan to combat antimicrobial resistance.* http://www.cdc.gov/drugresistance/actionplan/html/index.htm (accessed March 17, 2004).

U.S. Department of Agriculture. 2002. National organic program. http://www.ams.usda.gov/nop/Consumers/brochure.html (accessed March 17, 2004).

U.S. Department of Agriculture. Agricultural Research Service. 1998. USDA researchers create new product that reduces *Salmonella* in chickens. Press release. March 19.

U.S. Food and Drug Administration. Center for Veterinary Medicine. 1999. A proposed framework for evaluation and assuring the human safety of the microbial effects of antimicrobial new animal drugs intended for use in food-producing animals. Washington, D.C. February.

U.S. Food and Drug Administration. 2000. FDA/CVM proposes to withdraw poultry fluoroquinolones approval. CVM Update. Oct. 26. http://www.fda.gov/cvm/antimicrobial/FQWithdrawal.html (accessed March 17, 2004).

U.S. General Accounting Office. 1999. Food safety: The agricultural use of antibiotics and its implications for human health. Washington, D.C. April.

Witte, Wolfgang. 1997. Impact of antibiotic use in animal feeding on resistance of bacterial pathogens in humans. In Derek J. Chadwick and Jamie Goode, eds. *Antibiotic Resistance: Origins, Evolution, Selection and Spread.* New York: John Wiley and Sons.

World Health Organization. 2000. *Overcoming antimicrobial resistance.* http://www.who.int/infectious-disease-report/2000/ch3.htm (accessed March 17, 2004).

Yancey, Robert J. Jr. 2000. Presentation at 100th general meeting of the American Society for Microbiology. Los Angeles. May 22.

Zerno, Marcus. 2000. Presentation at 100th general meeting of the American Society for Microbiology. Los Angeles. May 23.

Zhang, Goulong, Christopher R. Ross, and Frank Blecha. 2000. Porcine antimicrobial peptides: New prospects for ancient molecules of host defense. *Veterinary Research* 31(3): 277–96.

2

Agricultural Antibiotics

Features of a Controversy

Brian Martin

Is using antibiotics in livestock and poultry problematic? Critics say that this practice is contributing to antibiotic resistance, with potential risks to human health. Defenders say the risk to humans is exaggerated and that the benefits outweigh the risks. Surely this disagreement can be sorted out in a straightforward fashion: just collect the scientific evidence and make a judgment. But, unfortunately, things are not this easy.

Social scientists have been studying what are called "scientific controversies" for quite some time (Collins 1981; Mazur 1981; Nelkin 1979). This includes debates about supersonic passenger aircraft, nuclear winter, genetic engineering, solar neutrinos, continental drift, greenhouse effect, cancer treatments, and microwave radiation. Some of these controversies largely occur between specialist scientists, such as between physicists who hold different views about gravity waves (Collins 1985). Sometimes specialist disputes spill out into the public arena, such as the controversy about cold fusion, which has implications for energy production (Simon 2002). And sometimes social implications are central to the dispute, as in controversies about pesticides and nuclear power. The antibiotics-in-farm-animals controversy fits in here.

When both scientific and social dimensions are involved, it is possible to say that a social controversy accompanies a scientific controversy (Engelhardt and Caplan 1987). But separating these two dimensions is not easy, and it may be more sensible to say that the scientific and social aspects are intertwined in a given controversy. What really animates a controversy is the connection between knowledge and power. A controversy is more than an intellectual disagreement because of the tight connection between scientific knowledge and power-related factors such as reputations, careers, positions of authority, profits, policies, and social control. One important implication is that scientific evidence, on its own, can never resolve the controversy. Evidence can always be disputed and theories are always open to revision, so disputes can persist so long as participants are willing to pursue them. Rather than thinking of evidence as the definitive means for resolving controversy, it is more useful to think of evidence as a tool—along with other tools, such as money, connections, authority, and eloquence—that can be deployed in an ongoing struggle.

Here I present a number of generalizations about scientific controversies, based on previous research. For each generalization, I give a few examples from other controversies—especially those, such as fluoridation, that I have studied in depth—and then offer an assessment of its relevance to the controversy about antibiotics in livestock and poultry. When reading the other chapters in this section, as well as throughout the book, it will be helpful to bear in mind these typical features of a scientific controversy as the authors stake out their positions.

A Range of Arguments

The arguments used in a typical scientific controversy fall into a range of categories, for example, scientific, ethical, economic, political, and procedural. In making sense of the controversy, it is often helpful to distinguish and classify the different types of arguments, recognizing that categories sometimes overlap.

In the long-running debate about fluoridation, which involves adding fluoride to public water supplies as a means to reduce tooth decay in children, participants make four main types of argument (Martin 1991). First, the benefits of fluoridation: advocates say that they are significant, whereas critics say that they are overstated. Second, the risks of fluoridation: advocates say no significant hazards exist; critics cite evidence about dental fluorosis, allergic and intolerance reactions, and potential genetic effects. Third, ethical considerations: advocates say fluoridation is ethical because it improves the dental health of children whose parents cannot afford dental treatment; critics say that it is unethical to medicate a population with an uncontrolled dose of fluoride. Fourth, decision making: advocates say that governments, advised by dental experts, should make decisions about fluoridation; critics say that the public should be directly involved in decision making.

In the controversy about the farm use of antibiotics, a central argument concerns human health: is antibiotic use in livestock and poultry leading to human resistance to antibiotics that are important for controlling pathogens? This includes various scientific arguments about the paths by which antibiotic resistance that developed in animals can be transferred to humans, the relative contribution of medical and animal use of antibiotics to the development of human resistance to antibiotics, and the effect of banning specific animal antibiotics. (See chapters by Salyers and Singer for two different scientific views about the risks to human health).

Another key argument concerns the economic benefit of using antibiotics in animals. This is linked to various scientific issues, such as questions concerning the effectiveness of the therapeutic, prophylactic, and growth-promoting uses of antibiotics. A lesser argument, somewhat behind the scenes, concerns who should make these decisions: farmers, governments, scientists, or someone else? Furthermore, whose advice should prevail? How should potential risks be judged?

Claims about human health are potent tools in public debate, which seems to be why critics of animal antibiotics highlight them: doing so puts supporters of these antibiotics on the defensive, arguing that the link to human health has not been sufficiently established. This is an example of the power-knowledge connection in controversy, with the agenda for scientific dispute in part set by what has saliency in the public arena.

Another sort of argument commonly used in controversies is to point to authorities—experts, professional associations, governments—that have taken positions in support of one's own. This can be called the argument from authority, which is not really an argument at all but an encouragement to defer to those particular authorities. Critics of agricultural antibiotics use the argument from authority when they point to European government regulations that ban the agricultural use of particular antibiotics such as avoparcin in the Netherlands, and when they refer to statements by professional bodies such as the American Medical Association (see Barlam in this volume). Those on the other side attempt to neutralize these endorsements by focusing on the evidence itself, for example, by pointing to differences between European and U.S. agriculture. (Singer, in this volume, points out these differences).

Endorsements sometimes follow one another in a bandwagon effect, with governments or professional bodies seeking guidance or drawing inspiration from each other. Nevertheless, it is possible for controversies to stabilize, with different countries or regions taking different positions. The fluoridation controversy has persisted for decades with high levels of fluoridation in most English-speaking countries and very little in Europe. Similarly, it is quite conceivable that a similar contrast between animal antibiotics policy and use in the United States and Europe could develop and persist. This is an example of how "the evidence" may be insufficient to resolve a controversy.

Coherent Arguments

In polarized debates, everyone lines up on one side or another, with hardly anyone left in the middle ground. In terms of arguments, this means that most partisans will use every possible point to support their position. The result is "coherent arguments," with views about different types of arguments lining up like iron filings in a magnetic field.

The fluoridation debate has been highly polarized. Proponents say that fluoridation has large benefits, no risks, is ethical, and that governments should make the decision on the advice of dental experts. Opponents take every contrary position possible. It is hard to find anyone who says, for example, that fluoridation has no risks but is unethical.

Debates are likely to become polarized when stakes are high. Stakes include money, commitment, prestige, and credibility. Key partisans often develop a psychological attachment to their position. To succeed, they draw on every bit of positive evidence and every doubt about contrary evidence. Those who adopt a middle position are likely to come under pressure to support one side or the other.

The antibiotics-in-farm-animals debate seems to be moderately polarized, judging by the way that most commentators line up with one set of arguments or another. Those on one side are likely to adopt a group of positions that emphasizes risks to human health, does not rate highly the economic benefits of antibiotics as growth promoters, and supports European governments that regulate against such antibiotic use. Those on the other side are likely to adopt a contrary group of arguments that emphasizes scientific uncertainties about the effect on human health, assumes significant economic benefits from using antibiotics to promote animal growth, and supports the U.S. regulatory system, which gives industry strong leverage over policy. Critics deploy or deride the scientific arguments, depending on their stance. For example, they see the European evidence as relevant—or not—to the United States, and they regard the causal pathways along which antibiotic resistance moves from animals to humans as either a cause for concern or as not sufficiently established.

Rarely does a commentator say that using antibiotics to promote growth in farm animals poses serious risks to human health but that farmers should be able to use antibiotics as they see fit, or, on the other hand, that human health risks are minimal but so are the economic benefits of animal antibiotics compared to the alternatives. According to the report of a 2001 colloquium held by the American Academy of Microbiology, the participants agreed that there was a "strong polarization of views" about the effects of agricultural antibiotics (Isaacson and Torrence 2001, 6).

Some scientists present themselves as neutral commentators, providing facts but not opinion. Reservoirs of Antibiotic Resistance (ROAR) describes itself as a "network dedicated to generating a new impetus worldwide for research on commensal bacteria as reservoirs of resistance that can be transferred to human pathogens." One of its two stated key objectives is to "act as the definitive source of information" on this topic. (http://www.tufts.edu/med/apua/ROAR/roarhome.htm; accessed May 24, 2004). Despite such worthy intentions, such scientists and groups are at risk of being drawn into the controversy when partisans on one side or the other, or both, draw on their material for campaign purposes. It is impossible to be perfectly neutral, because highlighting one fact rather than another can play into the hands of critics on one side of the debate. By focusing on antibiotic resistance, ROAR is likely to be more useful to critics of antibiotic use in agriculture. The more polarized the controversy, the less feasible it is to be a truly neutral commentator.

Alternatives

In a vociferous controversy, not only do the two sides become entrenched but so do the terms of the debate itself. In the debate about nuclear power, proponents were for nuclear power and opponents were against it—and both sides thought the question of nuclear power was central. A few proponents tried to broaden the terms of the debate to the issue of safety, to be achieved by such means as different reactor designs or underground construction, but they received little attention. Similarly, a few opponents tried to broaden the terms of the debate by questioning the need for centrally supplied electrical power, proposing energy efficiency plus decentralized solar and wind power as alternatives, but they too were a minority voice.

A prominent controversy can operate like a vortex, sucking nearby issues into its framework and subordinating them. Winning the debate becomes so important to participants that they lose sight of wider purposes. The fluoridation debate has drawn attention away from alternative methods of combating tooth decay, such as reducing sugar in the diet.

Use of agricultural antibiotics has become an entrenched practice through corporate investment, skill development, and psychological commitment. Changing such a practice would be difficult, even supposing that an alternative means became available to achieve equal weight gains at lower cost, with no loss in animals' health status. That is because switching to the alternative would mean that different companies would reap economic benefits, workers would have to learn different routines (and some might lose their jobs), and everyone involved would have to think in different ways. This is the lesson from experiences with other entrenched technologies, for example, military weapons systems (Morison 1966).

Though some alternatives to agricultural antibiotics are available or in development—see Mlot, in this volume—the critics are not unified in endorsing a particular alternative. This means that there is no alternative interest group—a company that stands to profit from a big new market—to challenge the entrenched practice.

If the purpose of livestock and poultry production is the highest quality meat, this seems compatible with restriction of antibiotics to sick animals. Even assuming the purpose is industry profits, then restriction of antibiotics across the industry might not be detrimental. However, these apparently rational assessments stand little chance in the face of an entrenched technology. (The pharmaceutical industry, however, has nothing to gain economically from restrictions on antibiotics, unless demand for a more profitable alternative is created.)

From the point of view of those concerned about how human health is affected by antibiotic resistance, the focus on farm animals may seem a distraction from reducing excess antibiotic use among humans, by far the greater source of antibiotic resistance. Also sidelined are more far-reaching alternatives, such as reducing the amount of meat that people eat, which would improve health in industrialized countries, reduce costs, and improve the environment. However, the meat-reduction alternative is far off the mainstream agenda and seldom mentioned in commentary on the antibiotics controversy.

The point here is not to endorse any particular alternative but rather to emphasize that controversies often have the effect of making the assumptions underlying the debate seem natural. It seems obvious to many participants that human health is the key issue in the use of animal antibiotics, but some people may well consider such issues as jobs or animal welfare to be of greater significance.

Partisans

A few high-profile partisans lead most controversies. This is especially true when the scientific content is significant. Individuals can become linchpins if they have some level of scientific capability and credibility, combined with a flair for powerful expression, public exposition, confrontation, and/or campaigning. Prominent examples are the scientist and science popularizer Carl Sagan and the physician Helen Caldicott in nuclear war–related debates. Such individuals may or may not be the central scientific figure; sometimes a charismatic personality can make up for scientific inadequacies.

The dynamics of debate helps create high-profile partisans. A person who makes a contribution—publishes a relevant paper, gives a talk, writes a popular article—is likely to be contacted by campaigners, invited to make further contributions, perhaps approached by the media. Such a person, if inclined, can become more actively involved, and in fact better qualified to do so, having received information, contacts, feedback, and encouragement. Those who develop a reputation may be asked to testify at hearings, speak at major meetings, or write an op-ed piece.

Is the debate about agricultural antibiotics led by high-profile partisans? One prominent Australian opponent is Peter Collignon, a microbiologist. Compared to the nuclear power or nuclear weapons debates at their peak this controversy is low key, not engaging all that many members of the public. Media coverage is not intense. But if the debate becomes more prominent, a few partisans will become more visible as carriers of the public debate.

Although a few individuals receive disproportionate attention, especially in the media, the people behind the scenes actually keep campaigns going by collecting and circulating information, building networks, organizing meetings, raising funds, and acting as contacts for the media. These individuals might be called the campaigners, who can range from public relations executives in a well-funded campaign to low-paid or volunteer activists in a grassroots campaign. Occasionally, these campaigners are also high-profile partisans, but often the labor is divided. Campaigners may be less visible, but they are driving forces in many controversies.

Support

Who supports one side or another in a controversy, and why? In a perfectly rational and compassionate world, an individual would study the issues and decide which side to support on the basis of evidence and logic, in the context of universal values such as justice and human welfare. A few individuals approach this ideal, but in practice hardly anyone has the time, expertise, character, and independence of mind to make this sort of judgment. So if we turn to the practical realities of controversies, it is possible to observe the influences that shape the decisions of most of those involved.

A group is likely to support one side in a controversy if it is in its interests to do so. This is straightforward: pesticide manufacturers support pesticides; automobile manufacturers support road building; doctors support medical intervention. If a controversy is associated with a product or practice, the group almost always lines up accordingly. In the debate about the benefits and risks of pesticides, pesticide manufacturers defend pesticides and criticize alternatives. In debates about transportation planning, automobile manufacturers defend cars and roads and do little to advocate bike paths or public transportation. In debates about cancer treatment, the medical profession supports surgery, radiation, and chemotherapy and criticizes unconventional therapies.

When an organization takes a stand, its members are likely to follow. When corporate executives support pesticides, most employees will as well, because doing so is in their personal interest, namely, their jobs, salaries, and peer support. Few group members take the time to carefully assess evidence and arguments on both sides. They simply follow cues from their superiors, perhaps taking note of materials prepared by their side.

The pharmaceutical and agricultural industries have developed interests in the regular use of animal antibiotics. Therefore it is entirely predictable that industry organizations and individuals will support this practice. For example, the Animal Health Institute, "representing manufacturers of animal health products" in the United States, features on its website (http://www.ahi.org) comments and articles criticizing the claim that animal and human antibiotic resistance are closely associated. On the other hand, public health workers, for whom the concerns of these industries are of no particular moment, are more likely to be critics of using antibiotics in animals. For example, the Union of Concerned Scientists, which has a long history of adopting public interest stands that challenge government or industry positions, issued a report titled *Hogging It!* (Mellon, Benbrook, and Benbrook 2001) that, among other things, criticizes work by the Animal Health Institute. The Union of Concerned Scientists features on its website (http://www.ucsusa.org/) "Myths and Realities About Antibiotic Resistance," in a question-and-answer format, that opposes most uses of antibiotics in farm animals.

Industry-funded researchers are likely to be supporters of antibiotics, whereas researchers with no ties to industry are less easy to predict. In a *Lancet* forum on antibiotic resistance, positions were predictable: all overtly industry-affiliated contributors supported industry use of antibiotics, whereas all critics of animal antibiotics were affiliated with universities or public health organizations (Singer et al. 2003).

The pattern of industry-related support for a scientific position is the most obvious aspect of the link between power and knowledge in a controversy, reflecting the adage that money speaks, even in science, though with the proviso that some scientists speak back. But this does not mean that arguments that serve powerful interests can be dismissed out of hand, only that these arguments warrant extra scrutiny.

Attacks

One of the less savory aspects of controversies is the exercise of power against opponents. For example, the nuclear industry has threatened, reprimanded, compulsorily transferred, demoted, dismissed, and black-listed employees who have exposed safety violations (Freeman 1981). The epidemiologist Thomas Mancuso was funded by the Atomic Energy Commission to do a study of the effects of low-level ionizing radiation; when he didn't come up with the findings that the commission preferred, it used a biased review process to withdraw his funding (Bross 1981, 217–22).

Attackers use whatever resources are at their disposal: rhetorical, personal, editorial, economic, and political. They may attack their opponents verbally, in overt abuse or through hard-to-trace rumors. When the opponent is a subordinate—such as when an employee exposes unwelcome data—attacks can take the form of ostracism, petty harassment, threats, or physical intimidation. Editors and referees can use their power over publication to block opponents' submissions. Some opponents lose their jobs or grants. In some cases, they may find themselves publicly denounced in the media.

These and other methods of attack—constituting what may be called "suppression of dissent"—seem to be especially prevalent when dissident experts provide support to a social movement that is challenging a powerful interest group, as in the cases of nuclear power, pesticides, and fluoridation (Martin 1999). Each side may attempt to attack the other, but often one side has a preponderance of resources. This is a stark example of how power can affect the search for and expression of knowledge.

In 2003, Dr. Ruth Hall, a leading researcher on antibiotic resistance, lost her job at the Commonwealth Scientific and Industrial Research Organization, the major Australian government research body. Commenting on this, Dr. Graeme Laver of the Australian National University was quoted as saying, "It did occur to me, I am afraid, that commercial pressures of some sort may have been responsible. We all know that Dr. Hall has made many statements on television, in the press and so on, that the practice of feeding antibiotics to livestock in order to promote growth might lead to antibiotic-resistant bacterial pathogens which would adversely affect human health" (Schwartz 2003). In this case, like many others, the evidence is insufficient to prove suppression of dissent, though it is compatible with such an interpretation. Subtle and deniable attacks are more effective than blatant ones, which can cause outrage.

Significant dissent is rare. Few people make public statements about their employer or about the viewpoint dominant among their colleagues. In the tobacco industry, for example, despite decades of covering up findings damaging to the industry (Glantz et al. 1996), very few employees ever spoke up. Therefore, it is reasonable to expect that few, if any, pharmaceutical industry employees will publicly voice criticism of the use of antibiotics, and that those who do will suffer reprisals (Abraham 1995). However, university scientists face less risk in making public comment, especially because neither side in the dispute has a preponderance of support.

Evidence

When a controversy involves science, a natural reaction is to say, "Let's collect some more evidence, and that will resolve the dispute." Controversies seldom conform to this logical approach. Indeed, new evidence often has no major effect on the dynamics of a controversy.

During the controversy about whether vitamin C can help in the treatment of cancer, a major study showed that the vitamin had no benefit to cancer patients. However, the scientists supporting vitamin C refused to accept the findings. Instead, they argued that the study was flawed in its method of choosing subjects and administering the vitamin (Richards 1991).

Evidence is not the "answer" to controversies for several reasons. Someone can always challenge the evidence: the results may be due to experimental flaws, misinterpretation, or chance variation. Each side in a dispute interprets the evidence through its own conceptual lens, typically dismissing contrary findings as inadequate or irrelevant and pouncing on favorable findings as significant or definitive. Partisans in controversies usually develop a strong psychological commitment to their position.

In the case of controversies with important social dimensions, new evidence is even less likely to be definitive because ethical, political, economic, or other dimensions to the issue remain contentious. In the case of the fluoridation debate, some opponents said that they would remain opposed even if fluoridation were completely safe, because it involved compulsory medication.

If the power-knowledge connection is central to scientific controversies, it is not surprising that knowledge alone is insufficient to transform the debate. Rather than thinking of evidence as a basis for resolving a controversy, it can be more useful to think of evidence as a tool that partisans use in their efforts to win support. Evidence is part of each side's "resource

tool kit," along with eminent endorsers, money, alliances, and commitment by key partisans.

Based on this assessment, we can predict that new evidence will not
greatly affect the debate about agricultural antibiotics. If new evidence
becomes available that supports one side, that side's partisans will strongly
declare its relevance and significance, but they are likely to be disappointed that the evidence has so little influence.

Closure

If evidence is insufficient to resolve a controversy, what does bring it
to an end? This is the issue of "closure" (Engelhardt and Caplan 1987).
Sometimes partisans on one side lose interest or energy; some retire or
die. Sometimes the weight of opinion is so one-sided that the weaker side
is marginalized into near invisibility.

Where social controversy is strong, its fate is often determined by
decisions by powerful groups, even though scientific issues remain unre-
solved. In the 1980s, debate raged about nuclear winter, a drastic reduction in atmospheric temperatures claimed to be likely after a nuclear war.
With the end of the Cold War in 1989, the entire issue dropped from sight,
although the scientific issues were never resolved.

The debate about agricultural antibiotics might reach closure in several ways. One is that governments ban the use of antibiotics as growth
promoters; the European Union is taking this approach in part. Another
is that key purchasers demand antibiotic-free meat. The June 2003 announcement by McDonald's that it would ban or discourage use of antibiotics by its meat suppliers was a major shift in the controversy. The scientific issues did not need to be resolved for McDonald's to make a decision
informed by its own interests, namely, enhancing its reputation as a
provider of food that has minimal negative effects on the environment and
health. A dramatic expansion of organic farming, which prohibits growth
promoters, could also help move the debate toward closure.

Another way that the controversy could end is by development of
alternative ways to promote animal growth. Indeed, a ban on antibiotic
growth promoters could stimulate investigation into alternatives, such as
probiotics, thus making the animal antibiotic debate irrelevant. It is also
possible to imagine scenarios in which animal antibiotic use becomes
more widely accepted, for example, as a result of introducing new animal

antibiotics that are unrelated to human antibiotics or as a result of a declining concern about antibiotic resistance in general because other issues take priority. Dissemination of animal or human disease vectors by terrorists could affect the debate in unpredictable ways.

The aim here is not to predict the future but to point out that what is called a scientific controversy can reach closure by a variety of means, including political and economic processes—another feature of the intertwining of power and knowledge in such controversies. But it is unwise to assume that a debate is gone forever. With changed circumstances, a moribund issue can pop up again, with renewed contention, for example, as a result of claims that a surge in human disease is related to animal antibiotic resistance.

Conclusion

Different scientific controversies have many similarities, though each has its own special characteristics. For those who are partisan participants in a controversy, studying the dynamics of related controversies may be helpful for picking up ideas about how to be more effective. Partisans typically believe implicitly that they are correct in their stances, so the main thing they need to learn is how to do better in their advocacy.

Outsiders, though, may not care which side "wins." They may just want to know how to make sense of the clash. When experts disagree, how can a nonexpert decide? Looking at general treatments of controversies can help to explain some recurring patterns.

Policy makers have a more urgent problem: what to do now. It is tempting to wait for more evidence, and scientists often advocate further research (Isaacson and Torrence 2001, 12–13 and chapters by Salyers and Singer, in this volume). But, as I described the process earlier, new evidence seldom resolves a controversy. In any case, policy has to address not only evidence but also the wider social dimensions of the issue. That means making value judgments, such as when benefits to one group cause risks to another. Antibiotics bring benefits to agricultural producers now, with a potential but unknown risk to people in the future. There is no purely scientific way to weigh competing claims.

Policy making is no more a neutral process than is the debate about antibiotic resistance, especially because policy makers are under pressure from various groups. Furthermore, to even speak of "policy makers" is to make assumptions about who makes policy: is it government agencies, legislatures, the market, elite scientists, or some form of direct public participation? Intertwined with scientific controversies are implicit assumptions, and sometimes overt debates, about how decisions should be made. It is anyone's prerogative to join the public debate about agricultural antibiotics. Studying the dynamics of scientific controversies can offer some hints for being a more effective participant.

Acknowledgments

I thank Jo Handelsman, Abby Kinchy, Daniel Kleinman, and two anonymous reviewers for helpful comments on a draft.

References

Abraham, John. 1995. *Science, politics and the pharmaceutical industry: Controversy and bias in drug regulation.* London: UCL Press.

Bross, Irwin D. J. 1981. *Scientific strategies to save your life.* New York: Marcel Dekker.

Collins, H. M., ed. 1981. Knowledge and controversy: Studies of modern natural science. *Social Studies of Science* 11:3–158.

Collins, H. M. 1985. *Changing order.* London: Sage.

Engelhardt, H. Tristram Jr., and Arthur L. Caplan, eds. 1987. *Scientific controversies: Case studies in the resolution and closure of disputes in science and technology.* Cambridge: Cambridge University Press.

Freeman, Leslie J. 1981. *Nuclear witnesses: Insiders speak out.* New York: W. W. Norton.

Glantz, Stanton A. et al. 1996. *The cigarette papers.* Berkeley: University of California Press.

Isaacson, Richard E., and Mary E. Torrence. 2001. *The role of antibiotics in agriculture.* Washington, D.C.: American Academy of Microbiology.

Martin, Brian. 1991. *Scientific knowledge in controversy: The social dynamics of the fluoridation debate.* Albany: State University of New York Press.

———. 1999. Suppression of dissent in science. *Research in Social Problems and Public Policy* 7:105–35.

Mazur, Allan. 1981. *The dynamics of technical controversy.* Washington, D.C.: Communications Press.

Mellon, Margaret, Charles Benbrook, and Karen Lutz Benbrook. 2001. *Hogging it! Estimates of antibiotic abuse in livestock.* Cambridge, Mass.: Union of Concerned Scientists Publications.

Morison, Elting E. 1966. *Men, machines, and modern times.* Cambridge, Mass.: MIT Press.

Nelkin, Dorothy, ed. 1979. *Controversy: Politics of technical decision.* Beverly Hills, Calif.: Sage.

Richards, Evelleen. 1991. *Vitamin C and cancer: Medicine or politics?* New York: St. Martin's Press.

Schwartz, Larry. 2003. No CSIRO place for top biologist. *The Age* (Melbourne, Australia), July 27.

Simon, Bart. 2002. *Undead science: Science studies and the afterlife of cold fusion.* New Brunswick, N.J.: Rutgers University Press.

Singer, Randall S. et al. 2003. Antibiotic resistance—The interplay between antibiotic use in animals and human beings. *Lancet Infectious Diseases* 3 (January): 47–51.

3

Agricultural Uses of Antibiotics
Evaluating Possible Safety Concerns
Abigail A. Salyers

Meat Producers and Purchasers Move Toward Reduced Use of Antibiotics

In June 2003, McDonald's made an announcement that sent shock waves through the livestock industry (McDonald's Establishes Global Policy 2003). Previously, with the exception of businesses that bought organic foods and a few specialty food outlets, meat producers and meat sellers had been content to look the other way when it came to the use of antibiotics in animal husbandry. McDonald's became the first large corporation to ask its meat suppliers to begin to phase out the use of antibiotics as a growth stimulant for food animals. Specifically, McDonald's asked its suppliers worldwide to "phase out, by the end of 2004, the use for the purpose of growth promotion of antibiotics that belong to classes of compounds approved for use in human medicine."

McDonald's was reacting to growing public concern about the widespread use of antibiotics in agriculture, most of it in animal husbandry. A spokesperson for the U.S. Food and Drug Administration (FDA) disagreed with the company's new policy, characterizing it as a "social policy rather than a scientific policy." McDonald's was not alone in wanting to move in this direction, for whatever reasons. Chicken producers such as Tyson's had already begun to reduce or eliminate antibiotics as growth promoters, and some smaller producers had been seeking to create niche markets by producing and advertising antibiotic-free meat.

All this occurred against the backdrop of a continuing examination of agricultural use of antibiotics by the FDA. The agency had already held hearings and seemed likely to hold more in the future. At issue was whether the FDA would withdraw authorization for the use of certain antibiotics in agriculture. The case that had drawn the most fire from consumer advocates was feeding the fluoroquinolone antibiotic enrofloxacin (Baytril) to chickens in their water to prevent development of an intestinal infection caused by *E. coli* that can decimate whole flocks of chickens.

Baytril was being used to prevent disease rather than simply promote the growth of chickens, but if a single infected bird was found, the antibiotic was administered to all chickens in the flock. Moreover, the U.S. Centers for Disease Control and Prevention had noted a rise in the incidence of fluoroquinolone-resistant *Salmonella* strains that cause infections in humans. This occurred shortly after chicken farmers began the widespread use of fluoroquinolones in animal husbandry. The possibility of a connection between the use of antibiotics in agriculture and a diminished ability to treat human infections in the future caused considerable concern. This concern spread to include not only Baytril but also those antibiotics being used to enhance the growth of farm animals such as chickens and pigs.

Since 1995 numerous conferences have been held on the agricultural use of antibiotics, and the European Union and U.S. government began supporting more research on how agricultural use of antibiotics might affect human health. It has become obvious that the question of how the use of antibiotics in agriculture could affect resistance patterns in human pathogens, and whether such an effect is in fact significant, is highly complex. It involves not just direct selection for resistance in bacteria that can colonize both humans and animals but also the potential transfer of resistance genes from such bacteria to human-specific bacterial pathogens. My purpose here is to provide an overview of some of the main scientific issues that have been raised and to assess the current status of the evidence available to support or refute concerns about agricultural use of antibiotics.

What Is the Safety Problem?

Before the Baytril controversy emerged, the aspect of agricultural use of antibiotics that most worried the FDA and consumers was antibiotic residues in foods, especially milk. The concern was that these residues would select for antibiotic-resistant human oral or intestinal bacteria that could later cause infection. Meat producers acted effectively to reduce antibiotic residues to a barely detectable level (see, for example, Dey, Thaler, and Gwozdz 2003, 391–404). Now concern has shifted from antibiotic residues in food to antibiotic-resistant bacteria in food.

The issue of resistant bacteria in food has proved to be far more complex than the issue of antibiotic residues in food. Not only are bacteria free-living organisms that can reproduce, but they are also capable of transferring their genes to other bacteria. Thus a bacterium need not be able to colonize a farm animal to benefit from resistance traits acquired by bacteria that do, if these two types of bacteria come into contact with each other in the human body. An added level of complexity arises because it still is not clear what types of antibiotic uses are most likely to select for resistant bacteria. That is, not all types of antibiotic use are equally dangerous. For example, most scientists now agree that long-term, low-dose antibiotic administration is more likely to select for resistant bacteria than short-term, high-dose administration because the former gives bacteria the opportunity to survive long enough to develop resistance by mutation or by acquiring DNA from other bacteria. Moreover, the longer regimens give bacteria the opportunity to acquire other mutations that may compensate for the deleterious effects of the mutation that makes the bacteria antibiotic resistant.

Types of Antibiotic Use in Animal Husbandry

In animal husbandry, antibiotics are used in three different ways (National Research Council 1999, 27–68). One of these uses is not controversial, at least so far—to treat sick animals. In this case, antibiotics are being used where they are clearly needed and are administered as short-term, high-dose treatments. A second use that has proved to be more controversial is the use of antibiotics to prevent infection (prophylaxis). The use of Baytril illustrates this. Clearly, if farmers are reasonably certain that disease-causing bacteria are present in their barns, it makes sense for them to try to avert any adverse consequences of this threat to their animals. However, some critics have pointed to the alternative, which is that better hygiene practices could eliminate the threat and thus make recourse to antibiotics unnecessary. Also, if the antibiotics being used for prophylaxis are also front-line drugs in the treatment of serious human infections, it is reasonable to ask whether the risk to human health is worth the economic advantage gained by the farmer.

The use of antibiotics to treat sick animals or to prevent disease both have parallels in human medicine, where antibiotics are used in much the same way. The third type of agricultural antibiotic use has no parallel in human medicine: the feeding of low concentrations of antibiotics to animals, particularly to pigs and chickens, to increase weight gain. The use of antibiotics as growth promoters is currently the most controversial. No one is sure why low levels of antibiotics, levels too low to kill most bacteria, have this growth-promoting effect. Also, the growth advantage is at most 5 percent. The animals do not necessarily become larger, but they gain weight more rapidly. Despite uncertainty about the mechanism of growth promotion and the seemingly marginal effects of this practice, farmers believe in it seriously enough to spend the money on medicated foods for their animals.

One hypothesis for how some antibiotics promote growth is that even low levels of antibiotics retard the number of some types of bacteria that stimulate the production and sloughing of animal intestinal cells (Gaskins, Collier, and Anderson, 2002). Mucosal cell turnover is one way that the human body prevents the invasion of bacteria that adhere to intestinal cells. Because the sloughed intestinal cells are dumped into the lumen of the intestine and excreted, a higher rate of intestinal cell sloughing can cause a significant loss of energy and protein. The antibiotics that are the best growth promoters, such as the tetracyclines, also have anti-inflammatory activity, which might also contribute to reducing the production and sloughing of intestinal cells, a process that has much in common with inflammation.

The Great Debate

As other chapters in this section have pointed out, the three main players in the debate about agricultural use of antibiotics are consumer advocates and environmental groups; farmers' organizations and pharmaceutical companies; and physicians. Scientists and government regulators have tried to play a neutral, objective role, as have academic research scientists working on the scientific basis of the risk-benefit aspect of agricultural antibiotic use. However, in my experience as a research scientist, it is not always easy to maintain a neutral view when working in an atmosphere that is charged with controversy. Scientists often find themselves being asked seemingly hostile political and economic questions after what they intended to be a rather dry scientific presentation.

To describe in detail the arguments of all groups interested in this issue is beyond the scope of this chapter, but there are some readily available sources for those who want more insight into the controversy. A number of consumer advocacy groups, including many groups that previously focused on environmental issues, have now joined the campaign against the widespread use of antibiotics in agriculture. The current main focus of these groups is the use of antibiotics as growth promoters. A recent book published by the Union of Concerned Scientists, *Hogging It!* gives an overview of the position of these groups (Mellon, Benbrook, and Benbrook, 2001). These are the groups that seem to have had the most influence on the new McDonald's policy, and they are the most active in the campaign to influence the FDA to rescind permission for certain types of antibiotic use in agriculture. (See Barlam's chapter in this volume for an analysis from the perspective of the UCS).

Farmers and the pharmaceutical industry defend the practice of using antibiotics as growth promoters just as staunchly. These groups point out that farmers have come to depend on antibiotics to maximize animal survival and rapid weight gain. They also point out that there is as yet no "smoking gun" that links agricultural use of antibiotics conclusively to antibiotic treatment failure in humans. For a more detailed summary of these arguments, consult the chapter by Singer in this volume or the websites of the Center for Veterinary Medicine (www.fda.gov/cvm/default .html) and the Animal Health Institute (www.ahi.org). Although the Center for Veterinary Medicine is a part of the FDA, it traditionally has been more closely allied philosophically with veterinarians and farmers' groups than with other groups with an interest in the issue.

Physicians have weighed in on both sides of the debate. Some think that those who campaign against agricultural use of antibiotics are missing the point, that consumer advocates, scientists, and policy makers ought to focus instead on misuse and abuse of antibiotics by physicians and their patients, which most agree is the primary force driving the evolution of antibiotic-resistant bacteria. Others are swayed by their concern about increasingly resistant bacteria, a concern that leads them to take the approach that it is better to err on the side of caution and drastically limit the use of antibiotics in animal husbandry. Reporters and the general public have tended to side with this latter view.

Independent scientists like myself who have worked to develop a scientific basis for making decisions about the potential risks of agricultural use of antibiotics are not members of this debate, in the sense that we try not to take sides or to let the political debate influence the outcome of our investigations. Yet, if scientists are to make an effective contribution, we need to be aware of what arguments are being made and how the various sides are likely to use our work. Scientists have also been put in the position of trying to help reporters understand the often bewildering scientific issues involved. In my experience, most reporters want to write accurate and balanced stories but rarely have the scientific training that could prepare them to unravel the confusingly tangled strands of the antibiotic controversy.

A good source of information on the scientific basis for evaluating claims about agricultural use of antibiotics is the website of the Reservoirs of Antibiotic Resistance Network at http://www.tufts.edu/med/apua/ROAR/roarhome.htm (accessed May 24, 2004). Stuart Levy of the Alliance for the Prudent Use of Antibiotics and I created this website to serve as a source of scientifically reliable information about antibiotic resistance in bacteria that are always present in the human or animal body but do not usually cause disease (commensals). Although normally innocuous, these bacteria can cause disease if they leave the environment where they are normally found (e.g., skin, mouth, intestine). Most postsurgical infections, for example are caused by human commensals. The commensals of humans and animals are at the center of the controversy about agricultural use of antibiotics.

The Amount of Antibiotics Used in Agriculture Is Still Uncertain

In assessing the risk that agricultural use of antibiotics might pose, it would be helpful to know, even approximately, how the amount of antibiotics used in agriculture compares to the amount used in human medicine. One frequently hears that the amount of antibiotics used in agriculture in the United States is comparable to the amount used in human medicine. In fact, this is only a guess because we have no solid, publicly available figures for the number of tons of antibiotics used each year in agriculture. The pharmaceutical industry has been reluctant to release its figures, presumably because it is reluctant to bring the volume of use to the attention of the public. Nonetheless, it is also true that what would constitute a reliable figure is difficult to define. For example, does one count the weight of active antibiotic or the weight of the antibiotic preparation, which may contain nontherapeutic additives? How does one take into account off-label use, that is, use of an antibiotic in a way that differs from the uses for which the antibiotic has been approved? These complications may explain why estimates of antibiotic use made by consumer advocates are far higher than those made by organizations such as the Center for Veterinary Medicine.

The Alliance for the Prudent Use of Antibiotics at Tufts University is sponsoring a study, carried out by a committee that includes representatives from industry, government, and academia. Its goal is to establish numbers with a solid scientific foundation that all sides can agree on. Despite the continuing uncertainty about the actual number of tons of antibiotics used in agriculture each year, everyone agrees that the number is very high and is likely to be comparable to, if not greater than, the amount used in human medicine.

What Is the Nature of the Risk?

As I mentioned earlier, the current concern about the widespread use of antibiotics in animal husbandry is that resistant bacteria selected by this use will enter the food supply and enter the intestines of consumers. This process can be divided roughly into three steps: selection of resistant bacteria in animals being fed antibiotics, appearance of these bacteria in meat offered for sale and consequent introduction of the bacteria into the human intestine, and the resulting consequences to consumers of having these bacteria pass through their intestinal tract.

Selection of Resistant Bacteria

Scientists have thoroughly documented the first step. Several reports document the presence of resistant bacteria in the intestinal tracts of animals being fed antibiotics (Salyers and McManus 2002). The Danish Zoonosis Institute (Danish Integrated Antimicrobial Resistance Monitoring and Research Programme [DANMAP]) has carried out one of the most extensive surveys of this sort. DANMAP monitors the use of antibiotics on Danish farms and reports the incidence of resistance in various sentinel bacteria such as *E. coli* and *Enterococcus* species (Aarestrup et al. 1998). DANMAP publishes yearly reports in booklet form.

Because it makes sense that antibiotic-resistant bacteria would be found where antibiotics are used regularly, it is easy to accept the results of such studies uncritically. The concept that heavy use of antibiotics increases the number of antibiotic-resistant bacteria is almost certainly correct in general, but there are some caveats (Salyers and McManus 2002).

First, surveys of resistant bacteria found in farm animals have begun fairly recently, so we essentially have no baseline to show whether this is an increasing problem. Second, some studies have failed to find a certain type of resistance on farms where such resistance would be expected. Conversely, resistant bacteria have been found on farms where antibiotics are used only occasionally, to treat sick animals. One explanation for this finding might be contamination by antibiotics in runoff from nearby farms that are using antibiotics, but as-yet-undiscovered factors might contribute to the emergence of antibiotic-resistant bacteria.

A third caveat is that the quantitative definition of *resistance* has not been established for all animal bacteria and all agriculturally used antibiotics. That is, the breakpoint for calling a bacterial strain resistant has not been established rigorously. This is being done, but in the absence of such guidelines one group might call a bacterial strain "resistant" to a particular antibiotic, whereas another group calls it "susceptible."

Introduction into the Human Intestine

Evidence for the second step, entry of the bacteria into the food supply and thence into the human intestine, is accumulating. Many reports have described isolation of antibiotic-resistant bacteria of animal origin from meat offered for sale to consumers (Salyers and McManus 2002). Perhaps the best evidence that bacteria on food enter the human intestine and survive long enough to be isolated from the colon come from an outbreak of *Salmonella* DT104 in the Netherlands and the vancomycin-resistant *Enterococcus* strains that originated on farms and ended up in the intestines of urban adults (Wegener 2003). In the case of the vancomycin-resistant enterococci, the farm origin of the bacteria is further supported by history: when the European Union banned the use of vancomycin in agriculture, the carriage of vancomycin-resistant bacteria dropped from about 10 percent to 4 percent. Although, in theory, cooking meat properly would prevent this sort of transmission, it is evident that this particular safety net has some fairly large holes in it. Thus, in considering whether bacteria on chicken, pork, or beef are likely to enter the human intestine, it is best to assume that from time to time some will make it through the preparation process. Because people consume food daily, the likelihood that such an event will occur eventually, even with the best preventative measures, rises considerably.

To prove the origin of a bacterium on food or isolated from the intestines of humans eating that food, it is necessary to do more than identify the species of the bacterium involved. Humans also carry many bacteria that are alleged to make this journey. The claim being made is not that these bacteria are introduced into the human intestine for the first time but rather that resistant forms are being introduced, thus increasing the number of resistant bacteria in the human colon.

Originally, in order to "fingerprint" a particular bacterium so that it could be traced from its origin to its destination, scientists used a method called serotyping. This approach uses antibodies to identify certain molecules on the surface of a bacterium. The multidrug-resistant strain of *Salmonella*, DT104, which has caused many outbreaks of salmonellosis, was identified in this way. Increasingly, however, scientists are turning to the use of DNA sequence information to "fingerprint" a particular bacterial strain (Salyers and McManus 2002).

Consequences to Consumers

The third step, adverse consequences arising from the presence of the ingested bacterium in one's intestine, is the most complicated to explain. The simplest consequence occurs when an intestinal pathogen like *Salmonella* or *Campylobacter* species enters the intestine and causes diarrhea. Some would argue that antibiotic resistance in such pathogens is irrelevant because antibiotics are not recommended for treatment of these self-limiting diseases. In a small percentage of the population, however, the pathogen can enter and spread throughout the body, causing a serious life-threatening infection. In such cases, antibiotic treatment is critical for saving the infected person's life.

Oddly enough, the bacteria that do not immediately cause infections present the greatest threat to the person carrying them. Most postsurgical infections and infections in cancer patients are caused by normally benign intestinal bacteria of the patient or of caregivers. These bacteria gain access to the tissue and bloodstream because of trauma to the intestine or breaches in the intestinal wall arising from other causes, such as chemotherapy. If many of a person's intestinal bacteria are resistant to antibiotics, infections caused by those bacteria may take longer to bring under control, resulting in longer hospital stays and greater damage to the patient. Put more directly, people who are carrying many resistant bacteria in their intestine could be considered to be ticking time bombs.

How would the passage of swallowed bacteria introduced on meat cause an increase of resistant bacteria in a person's intestine? This could happen in one of two ways. First, the incoming bacteria could colonize the person's colon. However, many scientists would agree that the likelihood that bacteria of animal origin would be able to compete effectively with native human colonic bacteria, even those that belong to the same species, is very small. Recent studies have suggested that this prediction is accurate. It is not surprising that incoming bacteria that are not adapted to the human body and have just passed through the acid barrier of the stomach might not do very well in a competition with bacteria that are already present in the colon. Also, many swallowed bacteria such as streptococci and staphylococci are not adapted to be successful colonizers of the human colon in the first place.

Bacteria that are merely passing through the intestine can, however, have an effect even if they do not persist in the colon: by transmitting their antibiotic resistance genes to resident bacteria. Resident bacteria may also transfer their resistance genes to the bacteria that are present transiently in the colon. After excretion, these bacteria can colonize other parts of the body such as the human mouth through the fecal-oral route. Even the most fastidiously clean people will transmit some bacteria on their hands from anus to mouth or skin. Some of the most feared bacterial pathogens, such as *Streptococcus pneumoniae* and *Staphylococcus aureus* are colonizers of the human throat and airways, whence they can take advantage of breaches in the defenses of the lungs or skin.

Bacteria can transfer genes by at least three processes: taking up DNA from the external environment (transformation), bacterial viruses that carry DNA from one bacterium to another (transduction), and direct transfer of bacterium to bacterium (conjugation). Conjugation is the type of transfer that causes the most concern because bacteria of different species and genera can transfer DNA to each other by this process. Transformation and transduction tend to be restricted to a particular species or subspecies. The process of conjugation takes as little as an hour, and even bacteria that are losing the competition with resident bacteria will be in the colon for days, plenty of time to pass on DNA to resident bacteria. Also, the very high concentrations of bacteria found in the human colon increase the probability that two bacteria would come into contact so that they could use conjugation to transmit DNA from one to the other.

In the laboratory, conjugative transfers between bacteria that have been recently isolated from natural settings generally occur at relatively low frequencies. Accordingly, scientists assumed initially that despite the high concentrations of bacteria found in the human colon, the likelihood that extensive DNA transfer would occur would be quite low. New evidence suggests that this assumption is incorrect. Evidence is accumulating that transfer of resistance genes by colonic bacteria, at least the numerically predominant populations, occurs more often than one might have expected (Salyers, Shoemaker, and Bonheyo 2002).

The results of a recent study of resistance gene transfer in the colon focused on *Bacteroides* species and indicated that transfer was indeed common (Shoemaker et al. 2001). *Bacteroides* species account for about 25 percent of the bacteria in the colon and are known to carry gene transfer elements that are transferred by conjugation. Because it would be unethical to feed antibiotic-resistant bacteria to human subjects in order to monitor the transfer of genes, this study used a different strategy. Researchers obtained from the human colon the DNA sequences of certain antibiotic-resistant genes found in various *Bacteroides* species. The rationale was that if two genes with virtually identical DNA sequences were found in different species, this result would be evidence that the gene had been transferred.

The study found that about 20 percent to 30 percent of *Bacteroides* isolated before 1970 carried *tetQ*, a tetracycline resistance gene, and none carried erythromycin resistance genes (*erm* genes). In the case of strains isolated after 1990, however, more than 80 percent carried *tetQ* and 15 percent to 20 percent carried *erm* genes. Genes found in different *Bacteroides* species had virtually identical DNA sequences, indicating that the rise in carriage of *tetQ* and the *erm* genes was the result of gene transfer. These genes proved to be carried on conjugative elements called conjugative transposons, so conjugation is likely to have been responsible for transfer. A limitation of this type of retrospective study is that although it provides evidence that transfer occurs, it does not determine the direction of transfer. Also, this type of study does not give quantitative information about the rate of transfer. An interesting aspect of the study's results was that the type of conjugative element found to be responsible for most of the gene transfer events is one whose transfer is stimulated considerably by the antibiotic tetracycline. Thus tetracycline use in the community may have been responsible for the extensive spread of the *tetQ* and *erm* genes that has occurred since 1970. Note, also, that tetracycline, which is a different type of antibiotic from the erythromycin family of antibiotics, could nonetheless stimulate the transfer of the erythromycin resistance (*erm*) genes. There has been a tendency to assume that one type of antibiotic was responsible for the spread and maintenance of resistance genes directed against only that antibiotic. Evidently, that is not necessarily the case.

This study provided support for the hypothesis that extensive resistance gene transfer has occurred among members of the genus *Bacteroides*. But unrelated bacteria, even of different genera or phyla, may also be able to exchange genes. Such gene transfer may be particularly important to establishment of resistance in the resident microflora if bacteria that are merely passing through the intestine can transmit their antibiotic resistance genes to resident bacteria. A combination of transfer of genes from transient members of the gut and across wide taxonomic divisions presents an ominous mechanism for spreading antibiotic resistance genes. More recent studies have provided support for the notion that such broad-range transfers are occurring in the human colon, transfers that may have involved such gram-positive pathogens as *Streptococcus pneumoniae* and *Bacteroides* species (Salyers, Shoemaker, and Bonheyo 2002; Wang et al. 2003).

Is Agricultural Use of Antibiotics an Environmental Issue?

As I stated earlier, farmers have done a good job of keeping antibiotic residues in human foods at very low levels. The same cannot be said of agricultural wastes that are dumped into the environment. Animal manure continues to be a serious environmental problem in a number of contexts, from noxious odors to contamination of groundwater with intestinal bacteria. Now contamination with antibiotics from farm runoff has become an issue (Campagnolo et al. 2002). Until recently, no one thought to measure antibiotic levels in streams and rivers, although it should have been obvious that the animal manure that accumulates on large farms would contain antibiotics. The assumption had been that antibiotics would not survive environmental conditions for very long. This seems to be the case for some, but others, like the fluoroquinolones, are proving to last longer in the environment than expected.

It is worth noting that one of the first reports of antibiotics in water sources, published in 1998 and reported in a *Science News* article that first brought greater attention to the problem (Raloff 1998), described water from a sewage treatment plant, not runoff from farm wastes. Thus farms are by no means the only sources of the antibiotics found in streams and rivers. The relative contribution of humans and animals as sources of excreted antibiotics is unknown. The consequences of low levels of antibiotics in the water supply are also uncertain. In theory, even low levels of antibiotics could disrupt bacterial populations normally found in environmental sites, but we do not yet know how susceptible these populations are.

Well before antibiotics were detected in water supplies, antibiotic-resistant bacteria had been found in many locations (McKeon, Calabrese, and Bissonnette 1995; Salyers and McManus 2002). Reports of antibiotic-resistant bacteria in a variety of environmental settings continue to appear. A number of these articles can be found in an annotated bibliography posted on the ROAR website mentioned earlier. A serious problem with assessing the effect of the agricultural use of antibiotics on the incidence of antibiotic-resistant bacteria in the environment is that virtually no information is available about the incidence of such bacteria before antibiotic use. Because soil bacteria produce many antibiotics, it should not be surprising to find resistant bacteria even in areas that have not been exposed to antibiotics, but the extent to which such bacteria occur naturally is not known. In this connection, it is noteworthy that in a study of human intestinal isolates obtained before 1970, we found that nearly one-third of the isolates were already resistant to tetracycline, although heavy use of tetracycline, both in human medicine and animal husbandry, began only in the late 1960s (Shoemaker et al. 2001).

Parting Thoughts

Clearly, the issue of agricultural use of antibiotics and its effect on human health is a complex one. It involves not only the movement of antibiotic-resistant bacteria but also of the resistance genes carried by these bacteria. To make matters more complicated, relatively little research has been done in this area until recently and there is still not much funding to support this type of research. Because of this, the amount of solid scientific evidence available is meager. Food purveyors such as McDonald's and Tyson Foods may well feel comfortable making decisions about how to handle antibiotic use in agriculture largely on the basis of public relations considerations. The FDA and USDA, however, have a responsibility to base decisions on rigorously validated scientific data. They are being challenged to make the best of the limited amount of data that they have to work with. It is never the case that the scientific basis for making regulatory decisions in a particular area is complete, with all the i's dotted and t's crossed, but in the case of agricultural use of antibiotics, the holes in the database are troublingly large. Also, it is not easy to explain the complex issues involved to the public, which ultimately must accept the outcome of decisions made in this area. More vigorous efforts need to be made to educate the public on these issues. The stakes for the average person are large. The loss of even some antibiotics as effective agents for treating and preventing life-threatening infections would seriously undermine the modern medicine that we have all come to take for granted. Yet preserving the economic viability of the food supply is also important. Meat production is the most obvious area that would be affected by a ban on the use of antibiotics in agriculture, but other areas of farming would also be affected (Salyers and McManus 2002). For example, antibiotics are sprayed on fruit trees to prevent bacterial infections that can wipe out an orchard. Also, antibiotic residues and antibiotic-resistant bacteria may well be found on plants that have been grown in fields irrigated with water that has been reclaimed from animal manure. Some balance must be struck between uncontrolled use of antibiotics and a complete ban on antibiotic use. Ideally, this balance will be determined on the basis of solid scientific data.

References

Aarestrup, F. M. et al. 1998. Surveillance of antimicrobial resistance in bacteria isolated from food animals to antimicrobial growth promoters and related therapeutic agents in Denmark. *APMIS* (*Acta Pathologica Microbiologica et Immunologica Scandinavica*) 106 (June): 606–22.

Campagnolo, E. R. et al. 2002. Antimicrobial residues in animal waste and water resources proximal to large-scale swine and poultry feeding operations. *Science of the Total Environment* 299 (November 1): 89–95.

Dey, B. P., A. Thaler, and F. Gwozdz. 2003. Analysis of microbiological screen test data for antimicrobial residues in food animals. *Journal of Environmental Science Health* 38 (3): 391–404.

Gaskins, H. R., C. T. Collier, and D. B. Anderson. 2002. Antibiotics as growth promotants: Mode of action. *Animal Biotechnology* 13 (1): 29–42.

McDonald's establishes global policy on growth-promoting antibiotic use. 2003. *Feedstuffs* 75 (June 23).

McKeon, D. M., J. P. Calabrese, and G. K. Bissonnette. 1995. Antibiotic resistant gram-negative bacteria in rural groundwater supplies. *Water Research* 29:1902–8.

Mellon, Margaret, Charles Benbrook, and Karen Lutz Benbrook. 2001. *Hogging it! Estimates of antibiotic abuse in livestock.* Cambridge, Mass.: Union of Concerned Scientists Publications.

National Research Council. 1999. *The use of drugs in food animals.* Washington, D.C.: National Academy Press.

Raloff, J. 1998. Drugged waters. *Science News* 153:187–89.

Salyers, A. A., and P. McManus. 2002. Possible impact on antibiotic resistance in human pathogens due to agricultural use of antibiotics. pp. 137–53. In *Antibiotic resistance and antibiotic development,* edited by D. Hughes and D. Andersson. New York: Harwood Academic.

Salyers, A. A., N. B. Shoemaker, and G. T. Bonheyo. 2002. The ecology of antibiotic resistance genes, pp. 1–17. In *Bacterial resistance to antimicrobials,* edited by Kim Lewis, Abigail Salyers, and Harry W. Taber. New York: Marcel Dekker.

Shoemaker, N. B., H. Vlamakis, K. Hayes, and A. A. Salyers. 2001. Evidence for extensive resistance gene transfer among *Bacteroides* spp. and among *Bacteroides* and other genera in the human colon. *Applied and Environmental Microbiology* 67:561–68.

Wang, Y., G. Wang, A. Shelby, N. B. Shoemaker, and A. A. Salyers. 2003. A newly discovered *Bacteroides* conjugative transposon CTnGERM1 contains genes also found in gram-positive bacteria. *Applied and Environmental Microbiology* 69 (8): 4595–4603.

Wegener, H. C. 2003. Antibiotics in animal feed and their role in resistance development. *Current Opinion in Microbiology.* 6 (October): 439–45.

4

Antibiotics in Animal Agriculture

An Ecosystem Dilemma

Randall S. Singer

Introduction

The increasing rate of development of bacterial resistance to antimicrobials has been well documented (Levin et al. 1998; Levy 1997; Levy 1998; Salyers and Amabile-Cuevas 1997). The development of this resistance has resulted in human and animal bacterial pathogens that are unresponsive to many forms of treatment currently available. Bacterial resistance to antibiotics has also become a high-profile, highly politicized health concern. With worries that the ability to treat bacterial infections might soon return to the pre-antibiotic era, researchers are attempting to identify new classes of antimicrobials as well as alternative therapies and prevention measures. When trying to solve this extremely complex problem of increasing rates of resistance, researchers naturally look for those activities that are major contributors and whose alteration or elimination would slow the loss of antibiotic efficacy. Antibiotic use is the major selection pressure influencing this situation, and because large amounts of antibiotics are used in animal agriculture, this practice has been labeled an overuse, and occasionally an abuse, of a valuable therapy for human health.

Most researchers and health care professionals will agree that the considerable misuse of antibiotics in human medicine has resulted in the high prevalence of resistant bacterial pathogens affecting human health today. However, there should be little argument that the use of antibiotics in animal agriculture provides a pressure that results in the selection of resistant bacteria and a pool of resistance genes. Because it is well established that bacteria, both resistant and susceptible, can be transferred from animals to humans and subsequently cause disease, the use of antibiotics in animals has some effect on human health. Several high-profile examples have highlighted the potential risk to human health posed by a resistant bacterial isolate that is transferred through the food chain (Fey et al. 2000; Molbak et al. 1999). Although these studies lacked the data to support the notion that antibiotic use in these animals *caused* the bacterial isolate in question to become resistant, it is clear that any use of antimicrobials will exert a selective pressure that can lead to the creation, amplification, dissemination, or persistence of antibiotic resistance mechanisms. This effect of antimicrobials occurs in many different populations and settings, including animals, plants, humans, and the environment. All these will add to the selective pressures exerted on bacteria; therefore, discerning *why* an organism has become resistant to an antibiotic or tracing the *origin* of an antibiotic resistance mechanism is extremely difficult.

Why has the use of antibiotics in animals been identified as a major component of the apparently escalating prevalence of resistant bacteria? If the use of antibiotics in animals has this effect, why are antibiotics still used in animal agriculture? Arguments about the use of antibiotics in animals persist not because the question is *whether* this use results in increasing resistance. The real questions that need resolution are *how much* does animal agricultural use affect resistance? *How significantly* does animal antibiotic use affect human health? *What are the repercussions* to human and animal health of the removal of antibiotics from animals? This chapter focuses on several specific topics that need clarification in order to resolve this debate. First, the importance of the background level of bacterial resistance must be recognized in the context of its relevance to assigning cause-and-effect relationships. Second, we must have a better understanding of risk in order to reach consensus on strategies that will minimize the human health effects associated with bacterial resistance. Third, different types of animal antimicrobial uses must be clarified in the context of the political and scientific policies that exist to regulate these uses. Finally, the ability to generalize the findings of individual studies to a national policy needs to be addressed in the context of U.S. agriculture.

For each of these topics, I will make suggestions about the types of studies that need to be conducted in order to make rational and scientifically sound decisions related to this complex issue.

Background Levels of Resistance

The current threat of antimicrobial resistance cannot be adequately determined strictly through a surveillance of bacterial pathogens. The majority of bacteria in the ecosystem of humans and animals are nonclinical and often exist in a commensal relationship with their host, meaning they are always present in the body but do not usually cause disease. Given the potential for resistance genes to move independently of their host bacterium, these commensal organisms may serve as reservoirs of antibiotic resistance genes if they are able to acquire these genes from and transfer them to pathogenic bacteria (Salyers and Shoemaker 1996; Salyers and Amabile-Cuevas 1997; Salyers 1995). Consequently, risks associated with antibiotic resistance might necessitate studies of the resistance gene as the unit of interest.

Resistance genes and resistant bacteria can exist in a diversity of ecosystems and have been detected in a wide variety of sources. For example, reports of resistant bacteria found in wildlife are becoming increasingly common, such as the recent documentation of resistant bacteria in raptors, sharks, and other wildlife (O'Rourke 2003). The discharge of wastewater from animal agricultural facilities (Aminov et al. 2002; Chee-Sanford et al. 2001), human sewage treatment plants (Kinde, Adelson, et al. 1997; Kinde, Read, et al. 1996), hospitals (Guardabassi et al. 1998; Reinthaler et al. 2003), and pharmaceutical plants (Guardabassi et al. 1998) has been associated with increased levels of pathogens as well as increasingly resistant and virulent organisms. Antibiotics are often discharged from these sites, and these antibiotics in the environment can act as a selection pressure further influencing the acquisition of resistance and virulence genes (Kummerer 2003; Kummerer and Henninger 2003). The global pool of resistance genes is extensive.

An assumption that is often made implicitly when investigating antibiotic resistance is that resistant bacteria detected on a farm, near a farm, in a processing plant/slaughterhouse, or on a food item obtained this resistance because of the use of antibiotics in animals. For example, a recent study (Chee-Sanford et al. 2001) investigated the presence of tetracycline resistance genes in groundwater near swine operations. These genes were found in the groundwater and matched genes found in the swine waste-water lagoons. The authors logically concluded that these genes existed in the environment as a direct result of agriculture. Unfortunately, the re-sults of other studies demonstrate the difficulty with making such sweep-ing conclusions. We have used a similar approach to detect tetracycline re-sistance genes in samples from livestock and wildlife, including wildlife in remote areas of the world with little contact with humans and no contact with human sources of antibiotics. In all these samples, we have detected tetracycline resistance genes that matched those found in the study by Chee-Sanford and colleagues. It is unlikely that the agricultural use of antibiotics caused these genes to disseminate around the globe and, in particular, into pristine ecosystems with no human influence.

Resistance genes and resistant bacteria, when found on agricultural premises or associated with animal-derived foods, can be resistant for rea-sons entirely independent of the use of antibiotics in animals. In fact, stud-ies have documented the introduction of foodborne bacterial pathogens onto agricultural premises from human sources, such as human waste-water plants (Kinde, Read, et al. 1996). Any use of antibiotics, including human, animal, and plant, as well as other naturally occurring compounds in the environment can amplify this background level of resistance. All these pressures must be accounted for in determining risks and in esti-mating the amount of risk attributable to animal antibiotic use.

Recently, Smith and colleagues (2002) used a mathematical model to assess the effect of agricultural antibiotic use on antibiotic resistance in human commensal bacteria. In this study, the bacterium, *Enterococcus,* was used as the model organism. The study concluded that under the as-sumptions of the model, the agricultural use of antibiotics does not nec-essarily increase the prevalence of resistant bacteria in humans beyond that which would be achieved because of the human use of the antibiotic. However, the appearance of this resistance would be hastened, and, under the specific enterococci example, the resistance levels would reach their asymptote (highest point) three years earlier when the antibiotic is used in animal agriculture.

A well-known statement about mathematical models goes like this: "All models are wrong, but some are useful." The problem with all mod-els is that they are completely dependent on the assumptions that drive them, as Barber, Miller, and McNamara (2003) discussed in a recent re-

Figure 4.1 How much of the antibiotic resistance observed on a farm is due to the antibiotic use on the farm?

view of the topic with respect to antimicrobial resistance and foodborne illness. A key assumption of the model by Smith and colleagues (2002) is that humans are exposed to new strains of resistant enterococci at a rate that is equal to the background rate of exposure plus the increased rate of exposure from agricultural antibiotic use. The model assumed that agricultural antibiotic use causes a thousandfold increase in human exposure to antibiotic-resistant bacteria.

One major problem with this model is its lack of a sensitivity analysis, which assesses the importance of the model parameters on the model outcome. A sensitivity analysis would reveal that the background level of resistance can have a huge influence on the model. In fact, by simply increasing the rate of background exposure, it is possible to lose three years of antibiotic efficacy without using the antibiotic in animal agriculture. Clearly, this parameter of background rate of exposure needs to be quantified more precisely. What is this background resistance, how do we measure it, and how might it influence a perceived cause-and-effect relationship between antibiotic use and antibiotic resistance?

Because of the large global pool of resistance genes and the use of antibiotics by humans for decades, determining the amount of human health risk attributable to animal antibiotic use is difficult. Well-designed longitudinal ecological studies are needed that incorporate a carefully selected spatial pattern of sampling. Tools such as geographic information systems

(GIS), which can relate multiple factors on spatial and temporal scales, need to be included to demonstrate the relationships between complex distributions of selection pressures and resistance genes. Ultimately, these studies would lead to an ecosystem-based transmission model, making it possible to extract a more accurate and precise estimation of the amount of risk to human health that can be attributed to the use of antibiotics in animals.

The Public Perception of Risk

Generally, the public has difficulty conceptualizing risk. Risk, as a probability, has a range from 0 to 100 percent; very few events have a probability at the limits. Most people have trouble deciding acceptable risk on their own and would prefer to know simply whether an event is "risky" or not, thus leaving the acceptable risk decision to someone else. Although most people feel more comfortable driving a car than flying in an airplane, decisions based on actual risk would suggest that these comfort levels should be reversed. A classic example of the public's misunderstanding and unrealistic expectation of risk relates to issues in food safety. In response to continued foodborne illnesses caused by organisms such as *Listeria, Salmonella, E. coli,* and *Campylobacter,* the government initiated several "zero tolerance" policies. For example, the U.S. Department of Agriculture's Food Safety and Inspection Service (FSIS) mandated that any ready-to-eat meat product containing detectable levels of *Listeria* or ground meat product containing *E. coli* O157:H7 be immediately recalled. Unfortunately, this policy has created the belief and expectation that meat products possess "zero risk," an entirely different and unobtainable goal.

The focus of the agricultural antibiotic use controversy should be to identify strategies that reduce the risk of resistant bacterial infections in people and in animals. To assess the ability of a strategy to reduce risk, we must be able to compare the amount of risk associated with an intervention versus the amount of risk associated with the status quo. This measure is frequently calculated as an attributable fraction (Barza and Travers 2002), but, unfortunately, the meaning of this measure is often misunderstood. A common epidemiologic definition of the attributable fraction is the amount of disease (or risk) that would be prevented if the factor, such as antibiotic use in animals, were removed (World Health Organization 2003b; Barza and Travers 2002). This definition assumes a causal relationship between the factor and the outcome. The attributable fraction, however, is a measure of association and, based on its calculation, makes no assumption about causality. As an example, studies that measure the frequency of resistance on farms that use antibiotics versus those farms that do not use antibiotics will give only an indication of the association between antibiotic use and resistance but will provide no information about whether the use of the antibiotic *caused* the resistance.

As I discussed earlier, resistant bacteria can exist on agricultural premises for many reasons that have nothing to do with the use of antibiotics in those facilities. Studies are needed that can fulfill specific causal criteria about the relationship between antibiotic uses and resistance and thus make it possible to estimate the strength of effect of antibiotic uses on resistance. The scientific community needs a better understanding of the differences between associations and causation and the ways in which each is measured. Finally, not all uses of antibiotics in agriculture are the same with respect to the amount of selection pressure that they exert, and, consequently, they cannot be lumped into a single input or risk. It is critical to understand the different uses of antibiotics in animals in order to develop rational and efficacious risk reduction strategies.

Antibiotic Use in Animal Agriculture

Antibiotics are used in animal agriculture in many ways, and many terms attempt to describe these uses. Unfortunately, considerable confusion exists regarding these uses and terms. For example, the American Veterinary Medical Association defines *therapeutic* as an antibiotic use that is intended for the treatment, control, or prevention of bacterial infections (American Veterinary Medical Association 1998). This definition thus relates to the treatment of disease in individual animals but also accounts for a population-based medicine. The term *metaphylaxis,* which refers to the medication of the entire herd or flock in order to treat sick animals and prevent infection in the remainder of the population, would thus be considered a therapeutic use. Prophylactic antibiotic use, which would include the use of antibiotics in the absence of disease but at a time when there is an expected increase in the incidence of disease, would also fit under this definition. On the contrary, the Union of Concerned Scientists and other groups use the term *nontherapeutic* to include all uses of antibiotics in the absence of disease, including use for growth promotion, metaphylaxis, and prophylaxis (Mellon, Benbrook, and Benbrook 2001).

Population-based Veterinary Medicine

Unfortunately, a population-based approach to medicine is difficult for people to comprehend. In many modern U.S. agricultural facilities, large numbers of animals are raised together, often in close confinement. When a disease outbreak begins under these conditions, its progression can be swift, with high morbidity and mortality. An antibiotic can be administered to the entire herd or flock in the face of the progressing disease. Some view this as a misuse or abuse of antibiotics because healthy animals are being treated. However, all animals in this facility can be considered to be exposed to the pathogen, and without treatment many of these exposed animals would eventually develop disease. Those that survive the outbreak may be affected in subclinical ways, such as being smaller, growing slower, or being generally less healthy than they would have been without the disease. For these animals, the antibiotic can prevent such untoward health effects.

An example of such a scenario would be an outbreak in broiler chickens of airsacculitis caused by *E. coli*. This respiratory disease in poultry can cause considerable morbidity and mortality in a very short time. The sight of such an outbreak can be dramatic, with hundreds to thousands of carcasses piled up in the broiler house in a single day. Because of the severe effect on animal health, the veterinarian will naturally want to treat this flock and prevent the disease from spreading to other birds in the house and other houses on the farm. One class of antibiotic therapy that is efficacious for treating outbreaks of *E. coli* airsacculitis is the fluoroquinolones. This use of fluoroquinolones in poultry has engendered a huge debate, a risk assessment by the U.S. Food and Drug Administration (FDA), multiple public hearings at the FDA, the proposed ban on fluoroquinolones for this purpose, and a subsequent lawsuit between the makers of veterinary fluoroquinolones and the FDA.

The scientific argument about the appropriateness of fluoroquinolone use in poultry is not the point of this discussion. It is clear from various experiments that fluoroquinolone resistance can develop rapidly in poultry-derived *Campylobacter* following the use of fluoroquinolones in poultry (Luo et al. 2003), but this does not necessarily mean that this use poses a high risk to human health. Rather, the point of this discussion is to clarify how and why an "important" human-use antibiotic is used in poultry. First, in the face of this airsacculitis outbreak, the only way to dose the entire flock with an antibiotic is through the feed or the water. In the case of the fluoroquinolone, therapy is administered in the water. This method of application has created considerable confusion because it has been equated with antibiotics used in feed for growth promotion, an obvious and understandable misconception. Second, it must be understood that fluoroquinolones and many other therapeutic antibiotics are expensive when they are used in food production animals at this population level. In the case of fluoroquinolones and airsacculitis, a threshold of morbidity and mortality must be reached before the veterinarian would prescribe the use of this antibiotic. Because of the decreased net return on the chickens in that flock, economics becomes the key factor underlying the decision to treat. Consequently, very few flocks actually receive fluoroquinolones, and some might not receive any antibiotic at all.

When considering the removal of an antibiotic from use, the debate typically focuses on the potential benefit of decreased resistance following the removal (World Health Organization 2003b). However, there may also be costs associated with the removal of the antibiotic. For example, repercussions could arise from forgoing an efficacious antibiotic during an airsacculitis outbreak or from removing from poultry production antibiotics that could reduce the overall incidence of airsacculitis. The spread of this disease produces high morbidity and mortality, and there is evidence that flocks of chickens showing signs of airsacculitis have lower weights, more fecal contamination, more errors at the time of processing, and consequently higher levels of *Campylobacter* spp. at processing (Russell 2003). Thus efforts to control airsacculitis might also be viewed as a means of producing a more robust and wholesome poultry supply and of reducing the risk of subsequent foodborne bacterial infection. We need studies that simultaneously evaluate the potential benefits and costs to human and animal health following the removal of antibiotics. It is critical to remember that the meat inspection process is intended to deliver a wholesome food supply; thus the presence of subclinical animal disease that could have a negative effect on human health is outside the scope of the current meat inspection process.

The Controversy Involving Antibiotics in Animal Feed

Growth promotion and enhancement of feed efficiency are by far the most controversial and misunderstood uses of antimicrobials in animals. The growth-promoting effect of adding antimicrobials to animal feeds was first described in the late 1940s by researchers studying the "animal protein factor," which was eventually identified as vitamin B12 (reviewed in Jones and Ricke 2003). However, the exact biological mechanism by which antimicrobials produce this effect remains incompletely understood. In the United States, some antimicrobials used in humans are identical to or in the same class as drugs currently approved for use as growth promoters in cattle, swine, and poultry.

Regardless of the precise amount of antibiotic that is used in animal feed, it is clear that large amounts of antibiotic are administered in this manner. This constant exposure of bacterial populations in the animal gastrointestinal tract to variable levels of antibiotic undoubtedly provides a persistent selection pressure in the population. In addition, much of these antibiotics and their metabolites may make their way into the environment (Kolpin et al. 2002) and continue to exert a selection pressure on environmental bacteria (Kummerer 2003; Kummerer and Henninger 2003). Few would argue against the notion that the use of antibiotics in this manner selects for resistance mechanisms or resistant bacteria in the animal population. However, the question that is being addressed in this case is whether the removal of these uses of antibiotics will aid human health through a decrease in resistance. Studies in the European Union, and primarily in Denmark and the Netherlands, have shown that following the removal of in-feed antibiotics that were labeled for growth promotion, the carriage of resistant fecal enterococci in animals (Aarestrup et al. 2001; Boerlin et al. 2001) and in people (Bruinsma et al. 2003) declined. It should be noted that many management changes in animal production were also made concurrent with these removals of antibiotics, and thus the observed changes in resistance may be the result of a combination of factors. The questions that must be asked based on these observations are whether human health has benefited and whether animal health has suffered from the removal of the antibiotics.

Antibiotic Use, Antibiotic Resistance, and Policy

Between 1995 and 2000, the European Union banned the use of the antibiotics avoparcin, bacitracin, spiramycin, tylosin, and virginiamycin for growth-promoting purposes. The public health perception was that these uses were having an effect on the increasing incidence of antibiotic resistance in human pathogens and therefore were increasing the risk to human health. Government authorities decided to implement control measures, including the removal of specific antibiotic uses in animals. Such an approach, often referred to as the "precautionary principle," aims to reduce potential risks through actions based on incomplete data. However, this approach can occasionally be misdirected, as well as counterproductive, by focusing resources away from more appropriate solutions (Starr 2003).

Since the ban of these antibiotics, the main effects that have been documented in humans have been a reduction in the prevalence of resistant enterococci isolated from the feces of animals on the farms and humans in the community (Aarestrup et al. 2001; Boerlin et al. 2001; Bruinsma et al. 2003). Hospital infections of clinically important vancomycin-resistant enterococci continue to increase in Europe, probably related to the increased use of vancomycin for the treatment of methicillin-resistant staphylococci. The amount of antibiotic that has been used therapeutically in these animal production systems has increased since the ban (Casewell et al. 2003), but the importance of this increase is controversial (Casewell et al. 2003; World Health Organization 2003a). Finally, the incidence of certain diseases in animals has increased, such as necrotic enteritis in broiler chickens (World Health Organization 2003a). As I described earlier, these changes in the incidence of animal disease could ultimately affect the safety of the food supply and the incidence of foodborne disease in humans. Regardless, it is clear that there are trade-offs to any strategy aimed at reducing the risks to human and animal health, and thus a sound policy for intervention will evaluate these risks before action is taken.

The "precautionary principle" is also presenting itself in the United States. Recently, two bills (S 1460 and HR 2932) were reintroduced in Congress under the title Preservation of Antibiotics for Medical Treatment Act of 2003. These bills are revisions of their initial 2002 form and aim to eliminate specific uses of antibiotics in animal agriculture.

A different approach has been adopted by many government regulatory authorities, industry associations, and other organizations that are using risk assessment methodology to determine and quantify the risks associated with the use of antibiotics in animal agriculture (Vose et al. 2001; Food and Drug Administration 2003). A risk assessment combines information about the consequence of an event with the probability that the event will occur. In this context, strategies for reducing risks can also be modeled relative to the baseline risk, thus identifying interventions that would be likely to affect the outcome.

The U.S. Food and Drug Administration's Center for Veterinary Medicine (FDA-CVM) has issued guidelines in its Regulatory Guidance Document 152 (Food and Drug Administration 2003) that can be used by animal drug sponsors to estimate the risk associated with the use of their antibiotic in animal populations. The goal of these guidelines is to demonstrate that the use of the antibiotic in animal agriculture does not compromise public health. Although the document is written for the preapproval of new animal drugs, it can also be useful for evaluating the risk of currently approved drugs for food animals.

Recently, a risk assessment model was developed under the FDA-CVM guidelines to assess the risk from the agricultural use of two macrolide antibiotics, tylosin and tilmicosin (Hurd et al. 2004). The model assessed all label claim uses of these antibiotics in the production of U.S. poultry, swine, and beef cattle. Using the guidance document, the hazard was defined as illness (1) caused by foodborne bacteria with a resistance gene; (2) attributed to a specified animal-derived meat commodity; and (3) treated with a human-use drug of the same class. Risk was defined as the probability of this hazard, combined with the consequence of treatment failure because of resistant *Campylobacter* spp. or *Enterococcus faecium*. This farm-to-patient risk assessment model demonstrated that the use of tylosin and tilmicosin in food animals presents a very low risk to human health. There are caveats to this model, however. First, the model does not address the selection pressures that antibiotics in the environment might have on bacteria. Thus the active metabolites of tylosin and tilmicosin that find their way into the environment could still present a potential selection pressure that influences antibiotic resistance. The model does not account for the potential transfer of genes between different types of bacteria; the model follows pathogenic bacteria through the food chain (as directed by the FDA-CVM document). Overall, the development of this transparent risk model provides a foundation on which to discuss the strategies available for reducing the risks to human and animal health.

Many cite the reported successes in the European Union as evidence that antibiotics used for growth-promoting purposes should be eliminated immediately (see Barlam in this volume, for example). For the reasons that I have described here, it should now be clear that the benefits to human health are debatable, in part due to the potential repercussions of this action. Another point that many fail to recognize is the applicability or generalizability of the European "experiment" to U.S. agriculture. The WHO document that addresses the effect of the termination of antibiotic growth promoters in Denmark (World Health Organization 2003a) is frequently cited as proof that these antibiotic uses can be eliminated in the United States without serious repercussions. However, the document clearly states that "under conditions similar to those found in Denmark, the use of antimicrobials for the sole purpose of growth promotion can be discontinued" (World Health Organization 2003a, 8). Denmark has a population less than 2 percent the size of the U.S. population and has an agricultural system very different from that in the United States. It is not clear that the interpreted results of the antibiotic removal in Denmark can be generalized to the United States. Studies are desperately needed that focus on the applicability of management strategies over varying spatial scales and across national and international boundaries.

In relation to antibiotic resistance, there is no reason to wait on the implementation of rational risk management policies. However, risk management solutions that are based more on a perceived benefit than on actual science may have more serious human health implications than initially anticipated. Sufficient data exist to begin developing and implementing these types of scientifically sound risk reduction strategies. Unfortunately, the politicization of the antibiotic use and antibiotic resistance issue has divided the groups that need to work together on this problem, which is clearly an ecosystem-based issue. To force a "precautionary principle" resolution through legal channels will only engender future mistrust and stubbornness. I hope that the scientific avenues that exist in the United States for implementing policy change will be used to redefine the ways in which antibiotics are used in animal agriculture in order to best serve human and animal health.

References

Aarestrup, F. M. et al. 2001. Effect of abolishment of the use of antimicrobial agents for growth promotion on occurence of antimicrobial resistance in fecal enterococci from food animals in Denmark. *Antimicrobial Agents and Chemotherapy* 45 (7): 2054–59.

American Veterinary Medical Association. 1998. Judicious therapeutic use of antimicrobials. Position statement. http://www.avma.org/scienact/jtua/jtua98.asp (accessed June 14, 2004).

Aminov, R. I. et al. 2002. Development, validation, and application of PCR primers for detection of tetracycline efflux genes of gram-negative bacteria. *Applied and Environmental Microbiology* 68 (4): 1786–93.

Barber, D. A., G. Y. Miller, and P. E. McNamara. 2003. Models of antimicrobial resistance and foodborne illness: examining assumptions and practical applications. *Journal of Food Protection* 66 (4): 700–9.

Barza, M. and K. Travers. 2002. Excess infections due to antimicrobial resistance: the "attributable fraction". *Clinical Infectious Diseases* 34 (Suppl 3): S126–S130.

Boerlin, P. et al. 2001. Antimicrobial growth promoter ban and resistance to macrolides and vancomycin in enterococci from pigs. *Journal of Clinical Microbiology* 39 (11): 4193–95.

Bruinsma, N. et al. 2003. Antibiotic use and the prevalence of antibiotic resistance in bacteria from healthy volunteers in the Dutch community. *Infection* 31 (1): 9–14.

Casewell, M. et al. 2003. The European ban on growth-promoting antibiotics and emerging consequences for human and animal health. *Journal of Antimicrobial Chemotherapy* 52 (2): 159–61.

Chee-Sanford, J. C. et al. 2001. Occurrence and diversity of tetracycline resistance genes in lagoons and groundwater underlying two swine production facilities. *Applied and Environmental Microbiology* 67 (4): 1494–1502.

Fey, Paul D. et al. 2000. Ceftriaxone-resistant *Salmonella* infection acquired by a child from cattle. *New England Journal of Medicine* 342 (17): 1242–49.

Food and Drug Administration. Center for Veterinary Medicine. 2003. Guidance for industry: Evaluating the safety of antimicrobial new animal drugs with regard to their microbiological effects on bacteria of human health concern. Doc. No. 152. October 23. Available at http://www.fda.gov/cvm/guidance/fguide152.pdf (accessed June 2, 2004).

Guardabassi, L., A. Petersen, J. E. Olsen, and A. Dalsgaard. 1998. Antibiotic resistance in *Acinetobacter* spp. isolated from sewers receiving waste effluent from a hospital and a pharmaceutical plant. *Applied and Environmental Microbiology* 64 (9): 3499–3502.

Hurd, H. S. et al. 2004. The public health consequences of macrolide use in food animals: a deterministic risk assessment. *Journal of Food Protection* 67 (5): 980–92.

Jones, F. T., and S. C. Ricke. 2003. Observations on the history of the development of antimicrobials and their use in poultry feeds. *Poultry Science* 82 (4): 613–17.

Kinde, H., M. Adelson, et al. 1997. Prevalence of *Salmonella* in municipal sewage treatment plant effluents in southern California. *Avian Diseases* 41 (2): 392–98.

Kinde, H., D. H. Read, et al. 1996. Sewage effluent: likely source of *Salmonella enteritidis*, phage type 4 infection in a commercial chicken layer flock in southern California. *Avian Diseases* 40 (3): 672–76.

Kolpin, D. W. et al., 2002. Pharmaceuticals, hormones, and other organic waste-water contaminants in U.S. streams, 1999–2000: a national reconnaissance. *Environmental Science and Technology* 36 (6): 1202–11.

Kummerer, K. 2003. Significance of antibiotics in the environment. *Journal of Antimicrobial Chemotherapy* 52 (1): 5–7.

Kummerer, K., and A. Henninger. 2003. Promoting resistance by the emission of antibiotics from hospitals and households into effluent. *Clinical Microbiology and Infection* 9 (12): 1203–14.

Levin, B. R. et al. 1998. Resistance to antimicrobial chemotherapy: A prescription for research and action. *American Journal of the Medical Sciences* 315 (2): 87–94.

Levy, S. B. 1997. Antibiotic resistance: An ecological imbalance, pp. 1–9. In *Antibiotic resistance: origins, evolution, selection, and spread*, edited by Derek J. Chadwick and Jamie Goode. *Ciba Foundation Symposium* 207. New York: John Wiley and Sons.

———. 1998. Multidrug resistance—a sign of the times. *New England Journal of Medicine* 338 (19): 1376–78.

Luo, N. et al. 2003. In vivo selection of *Campylobacter* isolates with high levels of fluoroquinolone resistance associated with gyrA mutations and the function of the CmeABC efflux pump. *Antimicrobial Agents and Chemotherapy* 47 (1): 390–94.

Mellon, Margaret, Charles Benbrook, and Karen Lutz Benbrook. 2001. *Hogging it! Estimates of antibiotic abuse in livestock.* Cambridge, Mass.: Union of Concerned Scientists Publications.

Molbak, K. et al. 1999. An outbreak of multidrug-resistant, quinolone-resistant *Salmonella enterica* serotype typhimurium DT104. *New England Journal of Medicine* 341 (19): 1420–25.

O'Rourke, K. 2003. Antimicrobial resistance in wildlife: It's making a bigger splash than you think. *Journal of the American Veterinary Medical Association* 223 (6): 756–57.

Reinthaler, F. F. et al. 2003. Antibiotic resistance of *E. coli* in sewage and sludge. *Water Researh* 37 (8): 1685–90.

Russell, S. M. 2003. The effect of airsacculitis on bird weights, uniformity, fecal contamination, processing errors, and populations of *Campylobacter* spp. and *Escherichia coli. Poultry Science* 82 (8): 1326–31.

Salyers, A. A. 1995. *Antibiotic resistance transfer in the mammalian intestinal tract: implications for human health, food safety and biotechnology.* New York: Springer-Verlag.

Salyers, A. A. and C. F. Amabile-Cuevas. 1997. Why are antibiotic resistance genes so resistant to elimination? *Antimicrobial Agents and Chemotherapy* 41 (11): 2321–25.

Salyers, A. A. and N. B. Shoemaker. 1996. Resistance gene transfer in anaerobes: New insights, new problems. *Clinical Infectious Diseases* 23 (Suppl 1): S36–S43.

Smith, D. L. et al. 2002. Animal antibiotic use has an early but important impact on the emergence of antibiotic resistance in human commensal bacteria. *Proceedings of the National Academy of Sciences U.S.A.* 99 (9): 6434–39.

Starr, C. 2003. The precautionary principle versus risk analysis. *Risk Analysis* 23 (1): 1–3.

Vose, D. et al. 2001. Antimicrobial resistance: risk analysis methodology for the potential impact on public health of antimicrobial resistant bacteria of animal origin. *Revue Scientifique et Technique* 20 (3): 811–27.

World Health Organization. 2003a. Impacts of antimicrobial growth promoter termination in Denmark. Report of a WHO international review panel, Foulum, Denmark. November 6–9, 2002. WHO/CDS/CPE/ZFK/2003.1.

———. 2003b. Joint FAO/OIE/WHO expert workshop on non-human antimicrobial usage and antimicrobial resistance: scientific assessment, Geneva, Switzerland, December 1–5.

5

The Impact of Antibiotic Use in Agriculture on Human Health and the Appropriate Public Policy Response

Tamar Barlam

The discovery of antibiotics was one of the most important advances in medical treatments. However, the prevalence of antibiotic-resistant infections is increasing. Although resistance to antibiotics will develop even with appropriate use,[1] the inappropriate prescribing by physicians and poor infection-control practices accelerate the problem. In addition, the growing scientific consensus is that antibiotic use in the production of food animals is an important factor.

As described in previous chapters, antibiotics are given in animal agriculture therapeutically to animals that are sick or those closely exposed to sick animals in a flock or herd. Antibiotics also are given nontherapeutically, often as a feed or water additive, to healthy animals for purposes of growth promotion (antibiotics are thought to improve feed efficiency, allowing an animal to grow more rapidly on less food) and disease prevention. Disease is anticipated because the animals are raised in crowded, often stressful, conditions and frequently become ill. Most antibiotics used in agriculture are given without the advice of a veterinarian and are available to farmers as medicated feed additives or over the counter.

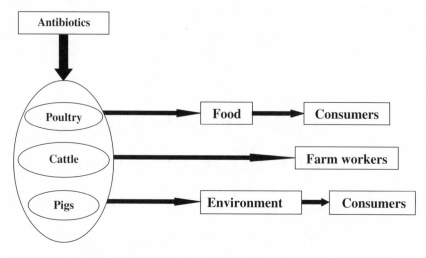

Figure 5.1   Possible routes of exposure/transfer.

The importance of agricultural antibiotic use in promoting antibiotic-resistant infections in people is controversial. There are three potential pathways by which antibiotic-resistant bacteria that originate in animals can reach people to cause disease (see figure 5.1). First, people who handle or eat improperly prepared food that is contaminated with bacteria can become ill. Second, the bacteria directly infect farmworkers, who may then in turn infect family members and eventually the community. And, finally, the bacteria and undigested antimicrobials can leach from farm waste into the environment and cause a public health risk. In this chapter, each of these pathways is discussed, followed by a discussion of an appropriate public health response.

The Food Pathway

Figure 5.2 describes the steps by which antibiotic use in livestock and poultry leads to antibiotic-resistant infections in people. First, antibiotics are administered to animals. In human medicine, it is generally believed that the greater the use, the greater the risk. Therefore, if large amounts of antibiotics are used in agriculture, that would likely pose a significant health risk. Second, the antibiotics select for antibiotic-resistant bacteria. Third, those antibiotic-resistant bacteria contaminate meat and poultry products. Finally, antibiotic-resistant bacteria appear in human bodies. To

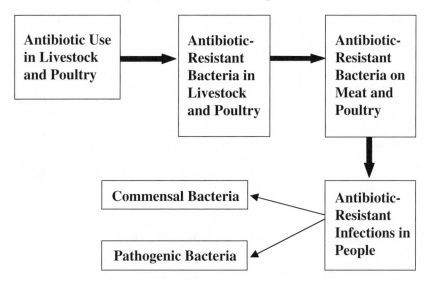

Figure 5.2 Antibiotic use in animal agriculture and its impact on human infections: the food pathway.

establish this final step, evidence of increasing antibiotic resistance in those bacteria most closely linked to animals must be found in affected humans. Although bacteria not carried by food may also acquire antibiotic resistance indirectly from antibiotic use in food animals through plasmid transfer of resistance elements, the focus of this chapter is on bacteria directly linked to animals—*Salmonella, Campylobacter, E. coli*, and enterococci.

Antibiotic Use in Food Animals

Three major sources have estimated the amount of antibiotics given to livestock and poultry (Animal Health Institute 2002; Institute of Medicine 1989; Mellon, Benbrook, and Benbrook 2001). In 1989, the Institute of Medicine estimated that almost thirty-two million pounds of antibiotics were used in animals in 1985 and approximately half of that amount was for growth promotion. The Animal Health Institute, which represents the pharmaceutical companies that produce antibiotics for animals, estimated that 18.2 million pounds were used in 2001 for treatment and prevention of disease and 3.7 million pounds were used for growth promotion. The estimate included all animals, including companion animals. The Union of Concerned Scientists (UCS) estimated that 24.6 million pounds of antibiotics were used nontherapeutically in food animals, and more than half

Table 5.1   Estimates of Antimicrobials Used Non-therapeutically in Animal Agriculture and Relative Importance in Treating Human Disease

Drug Classification	Cattle (pounds)	Swine (pounds)	Poultry (pounds)	Total (pounds)
Total Class I°	18,181	7,492	574,435	600,108
Total Class II°°	1,458,804	9,788,870	1,683,064	12,930,738
Class I and Class II as percentage of total	40 percent	95 percent	21 percent	55 percent
Total Class III°°°	2,216,032	552,234	8,278,427	11,046,693
Class III as percentage of total	60 percent	5 percent	79 percent	45 percent
Total all classes	3,693,017	10,348,596	10,535,926	24,577,539

Notes:
Adapted from Margaret Mellon, Charles Benbrook, and Karen Lutz Benbrook. *Estimates of Antimicrobial Abuse in Livestock.* UCS Publications, Cambridge, MA, 2001
°Class I drugs, as defined by report, are those used to treat human diseases, and have few or no alternatives, such as erythromycin or virginiamycin.
°°Class II drugs, as defined by report, are those used to treat human diseases, but alternatives exist, such as penicillins and tetracyclines.
°°°Class III drugs, as defined by report, are those not currently used to treat human disease, such as ionophores.

of the antibiotics were closely related or identical to antibiotics used by people (see table 5.1). In addition, UCS estimated that three million pounds were used in human medicine. Although the three estimates differ in many respects, all show that vast quantities of antimicrobial agents are being used in the agricultural sector.

Antibiotic-resistant Bacteria in Animals

In the second step of the food pathway selective pressure from the extensive veterinary use of antibiotics accelerates the emergence of antibiotic-resistant bacteria in animals. Tetracycline resistance increased among *Salmonella* isolates from food animals in the Netherlands when the drug was used in feed; resistance declined when this use was banned (van den Bogaard and Stobberingh 2000). The *S. typhimurium* strain DT104—a strain of *Salmonella* that is resistant to five different antibiotics—first circulated in cattle in the United Kingdom and then spread to animal populations throughout the world. Human disease consistently followed the emergence of this strain in animals in each discrete geographic area (Threlfall, Ward, Skinner, and Rowe 1997). When fluoroquinolone anti- biotics were allowed for poultry treatments, the rate of resistant *Campy- lobacter* rose from 0 percent to 14 percent in poultry in the Netherlands (Endtz et al. 1991) and from 0 percent to 10 percent in poultry in the United Kingdom. Levy and colleagues (1976) conducted a prospective study demonstrating that chickens given tetracycline-containing feed de- veloped tetracycline-resistant *E. coli* in their intestinal tracts. The prev- alence of antibiotic-resistant *E. coli* decreased when tetracycline was re- moved from the feed. In Europe, the use of the glycopeptide antibiotic avoparcin as a feed additive resulted in glycopeptide-resistant enterococci (GRE) in the feces of pigs and poultry (Bager, Masden, Christensen, and Aarestrup 1997). After avoparcin was banned as a feed additive for growth promotion, GRE decreased (Aarestrup et al. 2001).

Thus there is strong evidence that antibiotic use in livestock and poultry causes the emergence of antibiotic-resistant strains of bacteria in animals. Those bacteria in turn are found as contaminants on meat and poultry.

Antibiotic-resistant Bacteria on Food

Retail meats are commonly contaminated with antibiotic-resistant foodborne bacteria, which in turn can infect people. A recent study of a variety of ground meats and poultry in the Washington, D.C., area re- vealed that 20 percent of samples were contaminated with *Salmonella* strains, more than half of which were resistant to at least three antibiotics (White et al. 2001). Quinolone-resistant *Campylobacter* was identified on retail poultry products after the U.S. Food and Drug Administration (FDA) approved the use of fluoroquinolones for poultry treatments in 1995 (Smith et al. 1999). *Consumer Reports* ("Of Birds and Bacteria" 2002) tested 489 fresh whole broilers at supermarkets and health food stores in twenty-five cities. Forty-two percent of the broilers were con- taminated with *Campylobacter* and 12 percent with *Salmonella*. Ninety percent of the *Campylobacter* isolated were resistant to one or more anti- biotics, and 34 percent of the *Salmonella* strains were resistant to one or more antibiotics. McDonald and colleagues (2001) isolated streptogramin- resistant enterococci from 58 percent of retail chicken samples tested using selective media. Sørensen and colleagues (2001) found that ingested enterococci can colonize the intestinal tract of healthy volunteers for at least two weeks. Those organisms may transfer antibiotic resistance genes to other bacteria in the gut.

Antibiotic-resistant Infections in People

Antibiotic resistance in foodborne pathogens and commensal bacteria found in people is the final step in the pathway. Pathogens are bacteria that typically cause disease in humans, while commensals are bacteria that are always present in the body but do not usually cause disease.

PATHOGENS — *SALMONELLA* AND *CAMPYLOBACTER*   More than 1.4 million *Salmonella* infections occur in the United States each year (Mead et al. 1999), and the prevalence of antibiotic resistance in those infections is rising here and abroad (Glynn et al. 1998; Mølbak et al. 2002; van den Bogaard, Bruinsma, and Stobberingh 2000). *S. typhimurium* causes about one-quarter of *Salmonella* infections in this country; a 1995 survey revealed that 34 percent of infecting strains of *S. typhimurium* were resistant to at least five antibiotics. Most multidrug-resistant strains were of serotype DT104. Just fifteen years earlier, those highly resistant strains had accounted for only 0.6 percent of infections in humans (Glynn et al. 1998).

Outbreaks of DT104 infection have been directly associated with animal contact and ingestion of contaminated meats or unpasteurized dairy products (Cody et al. 1999; Villar et al. 1999; Tacket, Dominguez, Fisher, and Cohen 1985). For example, in 1999, an outbreak of infections with quinolone-resistant DT104 affected twenty-seven people, resulting in eleven hospitalizations and two deaths. The outbreak strain was identified in contaminated pork products that the patients had ingested and was further traced to two herds of swine (Mølbak et al. 1999).

Human infections resulting from other highly resistant types of salmonellae are emerging as well. Dunne and colleagues (2000) report that the prevalence of *Salmonella* infections with resistance to multiple antibiotics increased fivefold from 1996 to 1998. Two of the eleven patients interviewed for the study had visited a farm before their illness. The incidence of *Salmonella enterica* serotype newport, another common type of *Salmonella,* doubled from 1997 to 2001, and the prevalence of strains resistant to as many as nine antibiotics is increasing (Centers for Disease Control and Prevention 2002). Antibiotic-resistant *S.* newport infections have been epidemiologically linked to cattle exposure or beef ingestion.

In addition to *Salmonella* infections, *Campylobacter* causes more than a million foodborne infections in the United States each year. At least 60 percent of those infections are associated with contaminated poultry products. Resistance to the class of antibiotic called fluoroquinolones has increased among one type of *Campylobacter, C. jejuni,* since that class of drug became available for flockwide poultry treatments. In the Netherlands, quinolone-resistant *Campylobacter* infections in people were not seen until fluoroquinolones were used for poultry treatment in 1987 (Endtz et al. 1991). In the United States, fluoroquinolones were licensed for human use in the 1980s and for poultry treatment in 1995 and 1996. In a case-comparison study in Minnesota, the prevalence of quinolone-resistant *Campylobacter* infection increased from 1.3 percent in 1992 to 10.2 percent in 1998 (Smith et al. 1999). Molecular subtyping linked quinolone-resistant *Campylobacter* isolates from retail chicken products to strains in domestically acquired quinolone-resistant *Campylobacter* infections in humans. Thus there is strong evidence that the increase in antibiotic-resistant *Salmonella* and *Campylobacter* infections in humans is linked to the use of antibiotics in livestock.

COMMENSAL BACTERIA: ENTEROCOCCI AND *E. COLI*   Antibiotic resistance in enterococci, commensal bacteria that can sometimes cause disease particularly in patients debilitated by other medical problems, has been strongly linked to agricultural antibiotic use. Virginiamycin, a streptogramin antibiotic, has been used in agriculture for decades. Despite the lack of a related drug in human medicine until the recent approval of quinupristin-dalfopristin, about 1 percent of human fecal samples tested in a recent U.S. study contained streptogramin-resistant enterococci (McDonald et al. 2001). With increased use of quinupristin-dalfopristin, this intestinal reservoir of resistant enterococci may expand (Smith et al. 2002). In Europe, a 30 percent prevalence of streptogramin-resistant enterococci in the fecal flora of healthy adults declined to 12 percent within two years when use of virginiamycin was discontinued as an agricultural growth promoter (van den Bogaard et al. 2000); the prevalence is now under 2 percent. Use of avoparcin, a drug closely related to vancomycin and formerly used as growth promoter in Europe, caused the development of glycopeptide-resistant enterococci in the fecal flora of healthy people in the community; after the drug was banned, resistance rates declined (van den Bogaard and Stobberingh

2000). Belgian researchers recently reported a dramatic drop (from 5.7 percent to 0.8 percent) in the prevalence of vancomycin-resistant enterococci among hospitalized patients after avoparcin was banned. Human vancomycin use was stable (Ieven et al. 2001).

There is also evidence of increasing resistance in *E. coli* from antibiotic use in animals. *E. coli* are the major cause of urinary tract infections and can cause other serious infections in patients with other medical problems. In Spain, quinolone-resistant *E. coli* infections increased from 9 percent to 17 percent over five years, with a high prevalence in healthy children and adults that could not be explained by previous use of quinolone (Garau et al. 1999). Recently, Manges and colleagues (2001) found that a significant proportion of urinary tract infections caused by trimethoprim-sulfamethoxazole–resistant *E. coli* in three discrete geographic areas was clonally related. The most likely source of that clone was a common contaminated food source.

There is convincing scientific evidence that antibiotic-resistant bacteria spread from animals to people through food. Massive amounts of antibiotics are used in food animals. Antibiotic-resistant bacteria emerge in those animals, contaminate meat and poultry products, and result in a rise in antibiotic-resistant foodborne bacteria and infections in people.

The Farmworker Pathway

The second pathway by which antibiotic-resistant bacteria are transferred from animals to people is direct contact. Fey and colleagues (2000) report the case of a twelve-year-old boy with ceftriaxone-resistant *Salmonella* infection. Using pulsed-field gel electrophoresis and plasmid and β-lactamase analyses, the investigators demonstrated that the bacteria found in the child was indistinguishable from the bacteria obtained from a herd of sick cattle owned by his father.

Levy and colleagues (1976) conducted a prospective study demonstrating that chickens given tetracycline-containing feed developed tetracycline-resistant *E. coli* in their intestinal tracts and then transmitted those strains to a farmer and his family members. Ojeniyi (1989) reported that resistance patterns of *E. coli* isolates from workers at a research poultry farm matched those of isolates from poultry at the facility but differed from those of isolates from town residents and from birds outside the research facility. Nijsten and colleagues (1996) conducted a study in the

Netherlands, where fluoroquinolones were used in turkeys but not in pigs. These researchers found that only turkey farmers, and not pig farmers, had evidence of *E. coli* resistant to ciprofloxacin in fecal samples. The strains in farmers and turkeys were found to be identical by pulsed-field gel electrophoresis.

Nourseothricin is an antibiotic introduced in the 1980s in Europe as a growth promoter for pigs and is not used in human medicine. Within two years, a mobile resistance element with a nourseothricin resistance gene was found for the first time in *E. coli* isolated from the fecal flora of pig farmers, their family members, and nearby urban residents. The nourseothricin resistance gene was also found in *E. coli* strains causing community-acquired urinary tract infections. Later, the resistance gene was found in *Shigella,* an organism without an animal reservoir (Hummel, Tschape, and Witte 1986).

Thus there is clear evidence that antibiotic-resistant bacteria in farm animals can spread to farmworkers and from there into the community.

Environmental Pathway

The flow of antibiotic-resistant pathogens and undigested antimicrobials from farm waste into the environment is another potential pathway from medicated animal feed to human infections. The scientific data are limited, and no direct connection to human health has been demonstrated. However, several preliminary studies are of interest.

As much as 25 percent to 75 percent of antimicrobials administered to feedlot animals are excreted unaltered (Feinman and Matheson 1978, 372). The manure containing excreted antimicrobials is then applied to the land. Large holding pools, or lagoons, are also used for waste disposal at large farming operations. Seepage and runoff into watershed systems can occur, contaminating groundwater that is used for public water supplies and for drinking water in rural communities. In a 2001 study, tetracycline resistance genes were identified in the water of the two waste lagoons from which water samples were taken, in the underlying groundwater, and as far as 250 meters downstream (Chee-Sanford et al. 2001). In a recent report by the U.S. Geological Survey (Kolpin et al. 2002), 139 streams deemed susceptible to contamination, that is, downstream from intensive farming operations, were tested for pharmaceuticals, hormones, antibiotics used in human and/or veterinary medicine, and other organic contaminants, in thirty states. Low levels of one or more of the twenty-two antibiotics tested were found in almost half the waterways. In another study, Zahn (2001) looked at tylosin residues and tylosin-resistant bacteria in air from three mechanically ventilated swine confinements and found aerial transfer of both took place, suggesting a potential for wide dissemination of antibiotic residues and resistant bacteria. The retrospective study of Hamscher and colleagues (2003) lends credence to Zahn's results. Those researchers found the presence of up to five antibiotics in more than 90 percent of the air samples that had been collected for two decades from the same swine farm operation.

Public Policy Response

The data clearly illustrate that current antibiotic use in food animals can result in antibiotic-resistant infections in humans. The appropriate public policy response depends on a number of factors.

How Important Is the Public Health Problem?

Changing public policy in the face of scientific data gaps and industry opposition should be dependent upon the importance of the health problem of concern. The Task Force on Antimicrobial Resistance states in its action plan that "the extensive use of antimicrobial drugs has resulted in drug resistance that threatens to reverse the miracles of the last half century. Drug-resistant pathogens are a growing menace to all people, regardless of age, gender, or socioeconomic background" (2001, 9). The availability of effective antibiotics is especially essential for very young children, cancer patients, transplant patients, and others with immuno-suppressive conditions (Shea, Florini, and Barlam 2001; Shea 2003).

The strongest data that agricultural antibiotic use affects antibiotic resistance in bacteria that infect people are for food-borne organisms. *Salmonella* and *Campylobacter* cause millions of infections in the United States each year (Mead et al. 1999). *E. coli* causes approximately eight million urinary tract infections annually in this country (Manges et al. 2001), as well as other important diseases. *Enterococcus* is a serious hospital-acquired pathogen that particularly affects patients with underlying medical conditions. Although focusing on those bacteria alone is enough to justify action, many scientists believe that focus significantly underestimates the health risk (O'Brien 2002).

Is the Current Scientific Evidence Adequate to Take Action?

Many scientific, medical, and public health leaders are convinced that the scientific research is adequate to trigger a change in public policy. The Institute of Medicine (2003), the American Medical Association (2001), the Alliance for the Prudent Use of Antibiotics (FAAIR Scientific Advisory Panel 2002), the World Health Organization (2000), the American Public Health Association (2002), and others believe that the evidence has met the standard needed to take action. Based on existing data, the European Union has eliminated medically important antibiotics as feed additives and is proceeding to eliminate additional antimicrobial feed additives (European Commission 1998). Those respected groups have no obvious agenda other than protecting public health. There is not unanimity among scientists and physicians about the risk to human health caused by the agricultural use of medically important antibiotics, but the consensus is growing.

Those that question the adequacy of the science, although correctly identifying some data gaps that remain, are often stakeholders with a proprietary interest. More convincing support for withholding public action would come from impartial observers and through objective scientific evidence that demonstrates that widespread antibiotic use in agriculture does not pose public health harm. Instead, publications are often not peer reviewed or are sponsored by drug companies with a direct financial interest in the continued use of antibiotics in food animals (Bywater and Casewell 2000; Hurd et al. 2003; Doores et al. 2003). The call for more science is reasonable in theory, but how does one define the point at which the science is "enough"? (On this point see Martin chapter in this volume.) In addition, do those with and without a proprietary interest agree with that definition? History doesn't paint an encouraging view of industry's interpretation of when scientific certainty is adequate to take action to protect public health. At the turn of the century, despite the knowledge that milk was the vector for many infectious diseases, and the availability of a process—pasteurization—to make milk safe, the resistance to this technology was such that decades passed before all states required pasteurization (U.S. Centers for Disease Control and Prevention 1999). In another illustrative example, industry knew about the dangers of lead—it caused illness and death in workers—from the earliest days of gasoline production (Proctor 2003). However, the sale of leaded gasoline continued to be commonplace for decades. In 1973, the Environmental Protection Agency performed a definitive review of the issue and ordered a slow phasing out of the use of lead in gasoline. Industry fought for two years to stop this process (Lewis 1985).

Bearing this history in mind, if public health actions are delayed because of scientific uncertainty, impartial and knowledgeable scientists should define the next steps.

What Are The Challenges to Reducing Scientific Uncertainty and Further Defining Public Health Risk?

In the face of scientific uncertainty, one must consider whether the research to close the information gaps is feasible. If it is not, interventions to protect public health must be made despite uncertainty. In fact, there are many challenges to conducting research on the effect of agricultural antibiotic use on human health.

First, there has been a simultaneous use of most of the medically important antimicrobials in both animal agriculture and human medicine for decades. Second, the animal administered the drug is far removed from the person ingesting the meat, and there is little ability to trace the pathway from animal to infected person. Third, the resistance elements selected by the antibiotic use are often present on plasmids. Bacteria can pass plasmids containing resistance genes to other, unrelated bacterial species. Within the human gut, this is known to be commonplace (Shoemaker, Vlamakis, Hayes, and Salyers 2001) and markedly complicates tracing the path of resistance promoted by antibiotic use in animals. Fourth, for many reasons, resistant bacteria don't predictably decrease once a drug is removed. For example, resistance genes are often linked with other genes, such as those conferring resistance to other antibiotics (Aarestrup 2000) or to heavy metals (Summers 2002). If those antibiotics or metals are still used or are in the environment, selective pressure will continue. Fifth, molecular epidemiology techniques that have made the scientific evidence so convincing since the mid-1990s were not available until recently.

A final and important factor is that agribusiness and pharmaceutical companies control data that could help scientific study. Industry has the only reliable figures on amounts of antibiotics used, the stage of the life of the animal that received the antibiotic, and length of treatment. Surveillance of farm animals' enteric flora has been done only in limited experimental settings. The U.S. Department of Agriculture conducts on-farm projects only with the cooperation of agribusiness. That may undercut the scientific process. In Denmark, where detailed data on agricultural use of antibiotics are readily available and public (Danish Integrated Antimicrobial Resistance Monitoring 2002), studies have confirmed the effect of agricultural antibiotic use on animal and human colonization and infection with antibiotic-resistant bacteria.

For all of these reasons, it may be very difficult to establish the level of certainty about the human health effects of antibiotics used in agriculture that is called for by those who defend the use of antibiotics. Policymakers may need to act to protect public health despite uncertainty.

Could the Suggested Policy Changes Lead to Negative Effects That Offset Some of the Public Health Benefit?

Opponents of phasing out the nontherapeutic use of antibiotics in food animals have published several studies that indicate a negative effect on animal health and increased production costs if use of those drugs is discontinued (Casewell et al. 2003). Corporate-sponsored research has estimated that the cost of swine production in the United States would be minimally affected if nontherapeutic antimicrobial uses were banned in finishing pigs, but if removed throughout the life of the pig, the cost would increase to $4.50 per pig, which would mean a cost increase of about 2 percent to retail consumers (Hayes and Jensen 2003). However, estimates by an expert panel calculated that the cost of banning nontherapeutic use in agriculture of all antimicrobials would be only $4.84 to $9.72 per consumer per year (National Research Council 1999, 184).

In Denmark, feed additives containing antibiotics have been banned with a minimal effect on animal health (Emborg, Ersboll, Heuer, and Wegener 2001) and without causing undue financial stress to the farmers or an increase in retail meat prices (Wegener 2003). Recently, the World Health Organization convened an independent, multidisciplinary expert panel to review the Danish experience (2002). It found that overall antibiotic use had declined 54 percent since the ban (figure 5.3). Therapeutic uses had increased in swine but not in poultry; animal reservoirs of resistant bacteria had declined; and the economic effect was negligible in regard to poultry and 1 percent in overall production costs for swine (beef production is minimal in Denmark and was not examined). The report concluded that the use of medically important antimicrobials for the sole

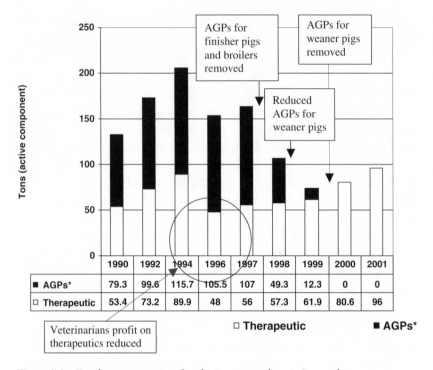

Figure 5.3 Trends in consumption of antibiotics in agriculture in Denmark.
Adapted from an unpublished figure by H. C. Wegener, using data from tables 3 and 9 of Danish Integrated Antimicrobial Resistance Monitoring and Research Programme, Danish Veterinary Institute, DANMAP 2002: Use of antimicrobial agents and occurrence of antimicrobial resistance in bacteria from food animals, foods and humans in Denmark (2002: 11–15).

purpose of growth promotion could be discontinued. In Switzerland, researchers studied the average therapy intensity for pigs after medicated feed additives were banned there and concluded that their results refute "the hypothesis that after the adoption of the ban on antimicrobial growth promoters, the amount of prescribed veterinary antibiotics increased" (Arnold, Gassner, Giger, and Zwahlen 2004). In other words, banning the use of antibiotics as growth promoters did not lead to an increase in other prescribed antibiotics to compensate.

Is the Current Regulatory Process Adequate?

The FDA approves new drugs for food animals if the manufacturer provides evidence not only that the drug is safe and efficacious for the animal but also that it does not endanger human health. When the majority of the medically important antimicrobials used in animal agriculture were approved, no one appreciated that their use could promote reservoirs of antibiotic-resistant bacteria and result in drug-resistant human infections. In an effort to address this issue, the FDA began the process of creating a guidance document for industry. A proposed framework was published in 1999 and finalized in October 2003 (FDA 2003a). Thus it took more than four years to produce a guidance document, and there is still no indication that the drugs of concern will be reviewed in a timely fashion. If the FDA did review one of those antibiotics and determined that the drug should be banned, a lengthy drug withdrawal process would then ensue. In response to a citizens' petition to the FDA that was led by the Center for Science in the Public Interest (1999), an FDA official said that it could take six to twenty years to remove a single antibiotic *if* the agency had the resources to do so (Sundlof 2001).

A recent example of the difficult and lengthy process of withdrawing a contested animal drug is enrofloxacin, a fluoroquinolone antibiotic approved for use in poultry. After approval, the FDA grew concerned when fluoroquinolone resistance in *Campylobacter* infections increased dramatically (FDA 2003b). After several years of conducting a risk assessment, in October 2000 the FDA proposed to withdraw the drug from use in poultry. In 2004, the process was continuing, and if the company that manufactures the drug continues to contest the withdrawal, the matter will not be resolved for many more years.

Is It Time to Take Action in the United States?

When there is no realistic expectation that the FDA can address the issue of antibiotic resistance in a timely manner, it is appropriate to seek alternative routes to protect public health. Legislation should be enacted that requires all sponsors of medically important antimicrobials used nontherapeutically in agriculture to resubmit their drug for approval, with the rationale that existing statutes at the time of original approval did not adequately address the effect on public health.[2] That strategy allows industry the opportunity to provide scientific data to the FDA demonstrating that nontherapeutic uses of its antibiotics do not harm human health. That is a science-based approach that would be protective of public health and maintain the integrity of the FDA process. It also would eliminate the possibility of lengthy and resource-intensive hearings that would bog down the FDA and prevent it from performing other essential functions.

Antibiotic-resistant infections increase morbidity and mortality, and health care costs (Office of Technology Assessment 1995; Phelps 1989; McGowan 2001; Rao, Jacobs, and Joyces 1988) and are an increasing public health problem. Agricultural use of medically important antibiotics contributes to that problem. In 2002, a multidisciplinary scientific advisory panel (FAAIR) concluded that use of antibiotics in animal agriculture posed a significant public health risk. The Institute of Medicine (2003) has called for an end to the use of medically important antibiotics for growth promotion in farm animals. Physicians and scientists have been willing to speak out in a variety of forums (Keep Antibiotics Working 2003). A scientific consensus is growing, and remedial action should be taken.

Notes

1. The terms antibiotics and antimicrobials are used interchangeably in this chapter.
2. As this book went to press, Congress was considering such legislation (Preservation of Antibiotics for Medical Treatment Act of 2003, introduced in the Senate—S 1460—and introduced in the House—HR 2932—in 2003).

References

Aarestrup, F. M. 2000. Characterization of glycopeptide-resistant *Enterococcus faecium* (GRE) from broilers and pigs in Denmark. *Journal of Clinical Microbiology* 38:2774–77.

Aarestrup, F. M. et al. 2001. Effect of abolishment of the use of antimicrobial agents for growth promotion on occurrence of antimicrobial resistance in fecal enterococci from food animals in Denmark. *Antimicrobial Agents and Chemotherapy* 45 (7): 2054–59.

American Medical Association. 2001. H-440.895 Antimicrobial use and resistance. http://www.ama-assn.org/apps/pf_online/pf_online (accessed June 4, 2004).

American Public Health Association. 2002. APHA Welcomes antibiotic resistance legislation designed to preserve the effectiveness of antibiotics for people. Press release, February 27. http://www.apha.org/news/press/2002/antibiotic-leg.htm (accessed June 4, 2004).

Animal Health Institute. 2002. Survey shows decline in antibiotic use in animals. Press release, September 29. http://www.ahi.org/mediaCenter/pressReleases/surveyShowsDecline.asp (accessed June 4, 2004).

Arnold, S., B. Gassner, T. Giger, and R. Zwahlen. 2004. Banning antimicrobial growth promoters in feedstuffs does not result in increased therapeutic use of antibiotics in medicated feed in farming. *Pharmacoepidemiology and Drug Safety.* 13: 323–31.

Bager, F., M. Masden, J. Christensen, and F. M. Aarestrup. 1997. Avoparcin used as a growth promoter is associated with the occurrence of vancomycin-resistant *Enterococcus faecium* on Danish poultry and pig farms. *Preventive Veterinary Medicine* 31 (1–2): 95–112.

Bywater, R., and M. Casewell. 2000. An assessment of the impact of antibiotic resistance in different bacterial species and of the contribution of animal sources to resistance in human infections. *Journal of Antimicrobial Chemotherapy* 46:643–45.

Casewell, M. et al. 2003. The European ban on growth-promoting antibiotics and emerging consequences for human and animal health. *Journal of Antimicrobial Chemotherapy* 52:159–61.

Center for Science in the Public Interest. 1999. Petition to rescind approvals of the subtherapeutic uses in livestock of antibiotics used in (or related to those used in) human medicine. Docket 99P-0485. http://www.cspinet.org/ar/petition_3_99.html (accessed June 4, 2004).

Chee-Sanford, J. C. et al. 2001. Occurrence and diversity of tetracycline resistance genes in lagoons and groundwater underlying two swine production facilities. *Applied and Environmental Microbiology* 67:1494–1502.

Cody, S. H. et al. 1999. Two outbreaks of multidrug-resistant *Salmonella* serotype typhimurium DT104 infections linked to raw-milk cheese in northern California. *Journal of the American Medical Association* 281:1895–10.

Danish Integrated Antimicrobial Resistance Monitoring and Research Programme. 2002. Danish Veterinary Institute. DANMAP 2002: Use of antimicrobial agents and occurrence of antimicrobial resistance in bacteria from food animals, foods and humans in Denmark. http://www.dfvf.dk/Files/Filer/Zoonosecentret/Publikationer/Danmap/Danmap_2002.pdf (accessed April 1, 2004).

Doores, S. et al. 2003. Low-level risk assessment for tylosin use in poultry on the treatment of human foodborne disease. Presentation Number C2–1502, Interscience Conference on Antimicrobial Agents and Chemotherapy Annual Meeting, Chicago, September 14–17.

Dunne, E. F. et al. 2000. Emergence of domestically acquired ceftriaxone-resistant *Salmonella* infections associated with AmpC β-lactamase. *Journal of the American Medical Association* 284:3151–56.

Emborg, H., A. K. Ersboll, O. E. Heuer, and H. C. Wegener. 2001. The effect of discontinuing the use of antimicrobial growth promoters on the productivity in the Danish broiler production. *Preventive Veterinary Medicine* 50:53–70.

Endtz, H. P. et al. 1991. Quinolone resistance in *Campylobacter* isolated from man and poultry following the introduction of fluoroquinolones in veterinary medicine. *Journal of Antimicrobial Chemotherapy* 27:199–208.

European Commission. 1998. Commission regulation of amending council directive 70/524/EEC concerning additives in feedingstuffs as regards withdrawal of the authorization of certain antibiotics. Document No. VI/7767/98. Brussels.

FAAIR Scientific Advisory Panel. Alliance for the Prudent Use of Antibiotics. 2002. The need to improve antimicrobial use in agriculture: Ecological and human health consequences. *Clinical Infectious Diseases* 34 (Suppl 3): S71–144.

Feinman, S. E. and J. C. Matheson. 1978. Draft environmental impact statement: Subtherapeutic antibacterial agents in animal feeds. Food and Drug Administration, Washington, D.C.

Fey, Paul D. et al. 2000. Ceftriaxone-resistant *Salmonella* infection acquired by a child from cattle. *New England Journal of Medicine* 342 (17): 1242–49.

Garau, J. et al. 1999. Emergence and dissemination of quinolone-resistant *Escherichia coli* in the community. *Antimicrobial Agents and Chemotherapy* 43:2736–41.

Glynn, M. K. et al. 1998. Emergence of multidrug-resistant *Salmonella enterica* serotype *typhimurium* DT104 infections in the United States. *New England Journal of Medicine* 338:1333–38.

Hamscher, G. et al. 2003. Antibiotics in dust originating from a pig-fattening farm: A new source of health hazard for farmers? *Environmental Health Perspectives* 111:1590–94.

Hayes, D. J., and H. H. Jensen. 2003. Lessons from the Danish ban on feed-grade antibiotics. Briefing paper 03-BP 41, Center for Agricultural and Rural Development, Iowa State University, Ames. http://www.card.iastate.edu/publications/synopsis.aspx?id=484 (accessed June 4, 2004).

Hummel, R., H. Tschape, and W. Witte. 1986. Spread of plasmid-mediated nourseothricin resistance due to antibiotic use in animal husbandry. *Journal of Basic Microbiology* 8:461–66.

Hurd, S. et al. 2003. Risk assessment of macrolide use in fed cattle on the treatment of human foodborne illness. Presentation Number K-1424, Interscience Conference on Antimicrobial Agents and Chemotherapy Annual Meeting, Chicago, September 14–17.

Ieven, G. et al. 2001. Significant decrease of GRE colonization rate in hospitalized patients after avoparcin ban in animals. Program and abstracts of the 41st Interscience Conference on Antimicrobial Agents and Chemotherapy, Chicago, December 16–19.

Institute of Medicine. 1989. *Human health risks with the subtherapeutic use of penicillin or tetracyclines in animal feed.* Washington, D.C.: National Academies Press.

———. 2003. *Microbial threats to health: Emergence, detection, and response.* Washington, D.C.: National Academies Press. http://www.nap.edu/books/030908864X/html/ (accessed June 4, 2004).

Keep Antibiotics Working. 2003. Safeguarding the effectiveness of existing antibiotics is essential. Paid advertisement. *Washington Times,* September 9. http://www.keepantibioticsworking.com/library/uploadedfiles/Letter_to_President_Bush.pdf (accessed June 4, 2004).

Kolpin, D. W. et al. 2002. Pharmaceuticals, hormones, and other organic waste-water contaminants in U.S. streams, 1999–2000: A national reconnaissance. *Environmental Science and Technology* 36:1202–11.

Levy, S. B., G. B. FitzGerald, and A. B. Macone. 1976. Changes in intestinal flora of farm personnel after introduction of a tetracycline-supplemented feed on a farm. *New England Journal of Medicine* 295:583–88.

Lewis, J. 1985. Lead poisoning: A historical perspective. *EPA Journal*, May. www .epa.gov/history/topics/perspect/lead.htm (June 4, 2004).

Manges, A. R. et al. 2001. Widespread distribution of urinary tract infections caused by a multidrug-resistant *Escherichia coli* clonal group. *New England Journal of Medicine* 345:1007–13.

McDonald, L. C. et al. 2001. Quinupristin-dalfopristin-resistant *Enterococcus faecium* on chicken and in human stool specimens. *New England Journal of Medicine* 345:1155–60.

McGowan, J. E. 2001. Economic impact of antimicrobial resistance. *Emerging Infectious Diseases* 7:286–92.

Mead, P. S. et al. 1999. Food-related illness and death in the United States. *Emerging Infectious Diseases* 5:607–25.

Mellon Margaret, Charles Benbrook, and Karen Lutz Benbrook. 2001. *Hogging It! Estimates of Antimicrobial Abuse in Livestock*. Cambridge, Mass.: Union of Concerned Scientists Publications.

Mølbak, K. et al. 1999. An outbreak of multidrug-resistant, quinolone-resistant *Salmonella enterica* serotype typhimurium DT104. *New England Journal of Medicine* 341:1420–25.

Mølbak, K., P. Gerner-Smidt, and H. C. Wegener. 2002. Increasing quinolone resistance in *Salmonella enterica* serotype enteritidis. *Emerging Infectious Diseases* 8: 514–15.

National Research Council. 1999. *The use of drugs in food animals: Benefits and risks*. Washington, D.C.: National Academy Press

Nijsten, R., N. London, A. van den Bogaard, and E. Stobberingh. 1996. Antibiotic resistance among *Escherichia coli* isolated from faecal samples of pig farmers and pigs. *Journal of Antimicrobial Chemotherapy* 37:1131–40.

O'Brien, T. F. 2002. Emergence, spread, and environmental effect of antimicrobial resistance: How use of an antimicrobial anywhere can increase resistance to any antimicrobial anywhere else. *Clinical Infectious Diseases* 34 (Suppl 3): S78–84.

Of birds and bacteria. 2003. *Consumer Reports*, January, pp. 24–28.

Office of Technology Assessment. 1995. *Impacts of Antibiotic-Resistant Bacteria (OTA-H-629)*. Washington, D.C.: Government Printing Office.

Ojeniyi, A. A. 1989. Direct transmission of *Escherichia coli* from poultry to humans. *Epidemiology and Infection* 103:513–22.

Phelps, C. E. 1989. Bug/drug resistance: sometimes less is more. *Medical Care* 27:194–203.

Proctor, R. N. 2003. Deceit and denial: The deadly politics of industrial pollution. *New England Journal of Medicine* 348:2696–97.

Rao, N., S. Jacobs and L. Joyce. 1988. Cost-effective eradication of an outbreak of methicillin-resistant *S. aureus* in a community teaching hospital. *Infection Control and Hospital Epidemiology* 9:255–60.

Shea, K. M. 2003. Antibiotic resistance: What is the impact of agricultural uses of antibiotics on children's health? *Pediatrics* 112:253–58.

Shea, K., K. Florini and T. Barlam. 2001. When wonder drugs don't work: How antibiotic resistance threatens children, seniors, and the medically vulnerable. Report, December 16. Environmental Defense, Washington D.C. http://www.environmentaldefense.org/documents/162_ABRreport%2Epdf (accessed June 4, 2004).

Shoemaker, N. B., H. Vlamakis, K. Hayes, and A. A. Salyers. 2001. Evidence for extensive resistance gene transfer among *Bacteroides* spp. and among *Bacteroides* and other genera in the human colon. *Applied and Environmental Microbiology* 67:561–68.

Smith, D. L. et al. 2002. Animal antibiotic use has an early but important impact on the emergence of antibiotic resistance in human commensal bacteria. *Proceedings of the National Academy of Sciences* 99:6434–39.

Smith, K. E. et al. 1999. Quinolone-resistant *Campylobacter jejuni* infections in Minnesota. *New England Journal of Medicine* 340:1525–32.

Sørensen, T. L. et al. 2001. Transient intestinal carriage after ingestion of antibiotic-resistant *Enterococcus faecium* from chicken and pork. *New England Journal of Medicine* 345:1161–66.

Summers, A. 2002. Generally overlooked fundamentals of bacterial genetics and ecology. *Clinical Infectious Diseases* 34 (Suppl 3): S85–S92.

Sundlof, Steven. Director, Center for Veterinary Medicine, U.S. Food and Drug Administration, to Center for Science in the Public Interest. 2001. http://www.fda.gov/cvm/efoi/citpet0485.pdf (accessed June 4, 2004).

Tacket, C. O., L. B. Dominguez, H. J. Fisher, and M. L. Cohen. 1985. An outbreak of multiple-drug-resistant *Salmonella enteritis* from raw milk. *Journal of the American Medical Association* 253:2058–60.

Task Force on Antimicrobial Resistance. 2001. A public health action plan to combat antimicrobial resistance. http://www.cdc.gov/drugresistance/actionplan/html/index.htm (accessed June 4, 2004).

Threlfall, E. J., L. R. Ward, J. A. Skinner, and B. Rowe. 1997. Increase in multiple antibiotic resistance in nontyphoidal salmonellas from humans in England and Wales: A comparison of data for 1994 and 1996. *Microbial Drug Resistance* 3:263–66.

U.S. Centers for Disease Control and Prevention. 1999. Achievements in public health, 1900–1999: Safer and healthier foods. *Morbidity and Mortality Weekly Report* 48:905–13.

U.S. Centers for Disease Control and Prevention. 2002. Outbreak of multidrug-resistant *Salmonella* Newport—United States, January–April 2002. *Morbidity and Mortality Weekly Report* 51:545–47.

U.S. Food and Drug Administration. Center for Veterinary Medicine. 1999. A proposed framework for evaluating and assuring the human safety of the microbial effects of antimicrobial new animal drugs intended for use in food-producing animals. *Federal Register* 64 (3): 887–88.

U.S. Food and Drug Administration. Center for Veterinary Medicine. 2003a. Guidance for Industry 152—evaluating the safety of antimicrobial new animal drugs with regard to their microbiological effects on bacteria of human health concern. October 23. http://www.fda.gov/cvm/guidance/fguide152 .pdf (accessed June 4, 2004).

U.S. Food and Drug Administration. Center for Veterinary Medicine. 2003b. Enrofloxacin for poultry: Opportunity for hearing. *Federal Register* 65 (October 31): 64594–65. http://www.fda.gov/OHRMS/DOCKETS/98fr/103100a.pdf (accessed June 4, 2004).

van den Bogaard, A. E., and E. E. Stobberingh. 2000. Epidemiology of resistance to antibiotics: Links between animals and humans. *International Journal of Antimicrobial Agents* 14:327–35.

van den Bogaard, A. E., N. Bruinsma, and E. E. Stobberingh. 2000. The effect of banning avoparcin on VRE carriage in the Netherlands. *Journal of Antimicrobial Chemotherapy* 46:146–48.

Villar, R. G. et al. 1999. Investigation of multidrug-resistant *Salmonella* serotype typhimurium DT104 infections linked to raw-milk cheese in Washington state. *Journal of the American Medical Association* 281:1811–16.

Wegener, H. C. 2003. Antibiotics in animal feed and their role in resistance development. *Current Opinions in Microbiology* 6:439–45.

White, D. G. et al. 2001. The isolation of antibiotic-resistant *Salmonella* from retail ground meats. *New England Journal of Medicine* 345:1147–54.

World Health Organization. 2000. WHO global principles for the containment of antimicrobial resistance in animals intended for food: Report of a WHO consultation. Geneva, June 5–9. http://www.who.int/salmsurv/en/Expertsreport-growthpromoterdenmark.pdf (accessed June 4, 2004).

———. 2002. Impacts of antimicrobial growth promoter termination in Denmark. Foulum, Denmark, November 6–9.

Zahn, J. A. 2001. Evidence for transfer of tylosin and tylosin-resistant bacteria in air from swine production facilities using subtherapeutic concentrations of tylan in feed. Published abstract, International Animal Agriculture and Food Science Conference, Indianapolis, Indiana, July 24–28.

PART 2
Genetically Modified Crops:
Global Issues

6

Genetic Modification and Gene Flow

An Overview

Allison A. Snow

Introduction

Debates about genetically modified crops have intensified around the world, often leading to legal, ethical, and international trade disputes. Starting in the late 1990s, the United States and a few other nations rapidly adopted genetically modified soybean, maize (corn), cotton, and canola, while many other countries chose a more cautious path or even outright rejection of genetically modified foods. To understand the interplay of the scientific and nonscientific issues at stake, it is useful to have a basic understanding of what genetically modified crops are and how easily their genes can disperse through human food supplies and the environment. In this essay, I highlight some direct and indirect effects of gene flow by focusing on genetically modified maize in two countries: the United States, where it is grown widely, and Mexico, which has a moratorium on planting this crop.

Why the Fuss about Genetic Modification?

Genetic modification, also called genetic engineering, refers to new methods of inserting genes into crop plants, fish, and other organisms to obtain useful traits (e.g., Snow 2003). In the past, the genes that were available for crop breeding were limited to those found in existing crop varieties and their sexually compatible wild relatives (such as teosinte, a wild relative of corn, as a source of maize genes). Now, however, scientists can take genes from unrelated organisms—viruses, bacteria, chickens, and so on—and insert them into plants to obtain valuable traits like resistance to pathogens.

Genes that have been artificially inserted into an organism's native genome are called transgenes. Transgenes from different species often perform their expected function in a completely unrelated species—for example, it is possible to move genes from fireflies or jellyfish into plants to make them glow in the dark (ISB 2004). Likewise, a bacterial gene that codes for an insect-killing protein has been inserted into maize to provide a built-in insecticide. Therefore, although the goals of genetic modification can be similar to conventional, old-fashioned breeding, this new method of obtaining genes for crop improvement is radically different (Gepts 2002; Snow 2003). A much wider range of commercially valuable products can be developed now, including plants that make new types of pharmaceutical and industrial chemicals. For the first time in history, scientists can create proprietary genes that can be added to any type of crop plant at will. Not surprisingly, this expensive and powerful technology has sparked a highly contested race among scientists, governments, and corporations to patent their new techniques for profit.

Meanwhile, many people have wondered whether there is something inherently risky about engineering the genomes of crop plants. To create a transgenic plant, specific DNA sequences are inserted into the native DNA of cultured plant cells, which are then regenerated into whole plants. The insertion process is somewhat random in that scientists cannot predict where a particular transgene will be inserted into the cell's genetic code, how many copies of the transgene will be transferred, or how well the transgene will function inside its new host. In maize, for example, a new transgene will be incorporated somewhere along the length of one of the crop's ten different chromosomes; later screening is done to ensure that a highly effective transgene is present.

It is fairly common for newly produced transgenic plants to exhibit abnormal growth forms and other unintended features. However, most researchers do not consider this to be a problem because they discard abnormal plants during the testing that takes place in the greenhouse and in the field. This kind of screening and selection is also carried out for novel *non*transgenic varieties that are produced from sexual crosses between species (and also may have unintended characteristics, at least at first). By the time a new variety is ready for commercialization, it has been pre-tested in various locations and growing seasons to ensure that it will not fall short of farmers' expectations.

Many expert scientific panels have concluded that the process of making transgenic plants is no more risky for human health or the environment than the methods that are used in conventional plant breeding (e.g., NRC 1987, 2000, 2002). The National Research Council and others note that, like other genes, transgenes are inherited, typically as dominant Mendelian traits, and it is the transgenes' products (e.g., plant-produced insecticides) that should be the subject of regulatory and scientific scrutiny, rather than the techniques used to obtain these products. This is why regulatory agencies examine transgenic crops on a case-by-case basis during evaluations. Current transgenic crops are expected to be safe for human health and the environment, although there is considerable debate about how the federal regulatory system should work and what types of worst case scenarios might occur if mistakes are made along the way (e.g, NRC 2002).

In the United States, three federal agencies must be consulted before genetically modified crops are used commercially (NRC 2002)—the U.S. Department of Agriculture, the U.S. Environmental Protection Agency, and the U.S. Food and Drug Administration. These agencies are responsible for making sure that any risks posed by genetically modified crops are no greater than the low level of risk associated with conventional breeding. The agencies evaluate transgenic crops with regard to (1) effects on human and animal health; (2) the potential for a transgenic crop to become a weed or to hybridize with a weed and become difficult to manage; and (3) the potential for a crop or its descendants to harm nontarget species such as pollinators, beneficial insects, and birds. These agencies give extra scrutiny to certain crops—among them, canola, squash, sunflower, rice, wheat, sorghum, turf grasses, and poplar trees—that can hybridize with wild relatives in the United States.

Transgenic crops are regulated and released by individual nations, but international differences in regulatory policies sometimes cause problems. By law, when regulatory agencies in the United States assess the risks of genetically modified crops, they do not consider potential health or environmental effects of transgenic plants in other countries, which typically have their own regulatory policies for genetically modified organisms. Also, the United States does not evaluate the effects of genetically modified crops on the genetic diversity of the crop or its wild relatives. Here, genetic diversity refers to the amount of genetic variation that is available for future crop breeding, as I discuss further below. So far, this has not been an issue because the United States is not a center of diversity for the four transgenic crops that are widely grown within its borders, namely, maize, soybean, cotton, or canola. Mexico, on the other hand, is where maize was first domesticated several thousand years ago. The Mexican countryside supports populations of the crop's wild ancestor (teosinte), as well as many locally bred maize varieties (landraces) that are grown by small-scale rural farmers. As I discuss later, this situation adds a new dimension to the science and politics of genetically modified maize in North America. As of this writing, Mexico has not approved the cultivation of genetically modified maize, although the Mexican government accepts and encourages imports of genetically modified grain from the United States.

Gene Flow and Its Consequences

When a transgenic crop is grown in farm fields, its genes are free to spread by means of pollen and seed dispersal. Gene flow is a natural process that occurs all the time among crops and their nearby relatives (Ellstrand 2003). In maize, for example, genes can be carried from one field to the next when the plant's male tassels shed their tiny, short-lived pollen grains into air currents. Most maize pollen falls within a few feet of its source, where the pollen may land on the female-receptive maize silks, but small amounts of pollen can fertilize seeds on plants that are more than a hundred feet away (Klein et al. 2003). Seeds that are fertilized by transgenic maize pollen bear a copy of the male parent's transgenes, which means that a neighboring *non*transgenic plant can produce transgenic seeds. Seeds also can have two copies of a given transgene if both of their parental plants are transgenic, or they can have one copy that is inherited from the seed's "mother" rather than the "father." (Typically, one copy of a transgene is enough to confer a new trait, such as herbicide resistance.) Transgenes that are present in seeds can be carried over many hundreds or thousands of miles because seeds are transported so extensively by people, wind, animals, and water. When genetically modified maize is shipped to other countries as grain for human food or animal feed, the seeds can germinate and grow if they are planted by farmers.

Given that gene flow is ubiquitous, it is not surprising that transgenes have been detected in places where they were not intended to spread, including certified, nontransgenic seed sources (Beckie et al. 2003; Mellon and Rissler 2004), organic food (Partridge and Murphy 2004), and maize fields in remote regions of Mexico (Quist and Chapela 2001, 2002). To some people, this represents a type of genetic "pollution" or "contamination." This perception is relevant to certain legal and trade issues, organic farmers' concerns, and ethical considerations regarding freedom of choice and informed consent on the part of farmers and consumers. However, from a purely scientific perspective, it is important to ask whether "contamination" from transgenes could have any unwanted biological consequences for humans or the environment, especially when the transgenes are present at very low levels. If gene flow really causes biological contamination, it must have unwanted biological consequences. To further focus this question, it is essential to ask how transgenes might differ from other types of crop genes that undoubtedly disperse in the same way, having done so for centuries and largely gone unnoticed. Some of the confusion and controversy surrounding genetically modified crops could be reduced if more people understood these underlying scientific questions.

Unfortunately, because genetically modified crops are so new, and because their development has been dominated by private industry, scientific studies of the biological effects of gene flow from transgenic crops are meager (e.g., Wolfenbarger and Phifer 2001; Dalton 2002; Pilson and Prendeville in press; Snow et al. 2004). Much more is known about where transgenes can spread than about the unintended consequences of gene flow (e.g., Snow 2002). Although this lack of knowledge clearly adds uncertainty to discussions about potential risks and benefits of transgenic crops, it is possible to make several predictions about these consequences with regard to specific examples of transgenic crops, as I discuss later. The complex and important topic of how transgenic crops could affect human and animal health is beyond the scope of this essay, but interested readers should look for the reviews by the Royal Society of Canada (2001), Haslberger (2003), the World Health Organization (2003), and the National Research Council (2004).

Genetically Modified Maize and Gene Flow in the United States

Because maize is an outcrossing, open-pollinated crop—that is, each plant disperses large amounts of pollen into the air to "sire" seeds on other plants—transgenes in maize are expected to disperse widely by means of both pollen and seeds. To illustrate how the biological consequences of gene flow can be evaluated, we can compare the example of genetically modified maize in the United States and in Mexico. First, we will consider genetically modified maize in the United States, which is the most likely source of genetically modified maize in Mexico.

Two major kinds of maize have been deregulated and commercialized in the United States—varieties with transgenic resistance to certain insect pests (these are known as Bt maize varieties), and varieties with transgenic resistance to certain herbicides (namely, glyphosate or glufosinate). These transgenic traits are not expected to cause significant environmental (or health) problems in the United States, based on scientific studies and general predictions. For example, Bt proteins are toxic only to narrow categories of insects, such as moths and butterflies. They do not appear to harm other species such as bees, birds, or humans because these proteins are easily digested and they are not concentrated in food chains. In fact, organic farmers often use Bt sprays on their crops because the compounds break down quickly and affect only targeted insect pests.

A few ecological studies have focused on Bt plants (e.g., Snow et al. 2003). A series of recent studies (see NRC 2002) has resolved questions about toxic effects of wind-dispersed Bt maize pollen on monarch butterflies. These investigators found that currently grown genetically modified maize varieties are not likely to harm monarch butterfly caterpillars, which rarely ingest large volumes of maize pollen. Also, although the plants' insecticidal Bt proteins can persist in the soil for several months, scientists found that these proteins are not likely to harm earthworms or other soil organisms (Saxena and Stotzky 2001).

Future investigations may reveal unanticipated problems with insect resistant or herbicide resistant maize in the United States, but so far their environmental effects appear to be minimal in comparison with the environmental effects of growing nontransgenic maize. In regions where pesticides are used frequently, such as on sweet corn, genetically modified maize allows farmers the option of reducing their pesticide application.

The economic and environmental benefits of Bt crops have been recognized by the Environmental Protection Agency, which has established policies to help delay the time when target insect pests evolve resistance to Bt crops (NRC 2000).

To summarize the situation in the United States, currently grown types of genetically modified maize are not expected to cause health or environmental problems. This biosafety evaluation is made easier because the maize cannot revert being a weedy form of the crop, as occurs in canola, for example, and maize has no sexually compatible wild relatives that could acquire new transgenic traits from the crop. Also, U.S. farmers generally do not save maize seed from one year to the next because they prefer high-yielding, modern varieties, which are sold anew each year as first-generation hybrids (each variety is a hybrid of two inbred maize lines, resulting in "hybrid vigor"). This means that transgenes could be "recalled," if this should ever become necessary, by excluding them from the new batches of hybrid seed that are sold to farmers in the spring.

Genetically Modified Maize and Gene Flow in Mexico

The biological, economic, and cultural aspects of maize cultivation are very different in Mexico. Not only is there a common wild relative, teosinte, but about two-thirds of the maize grown in Mexico consists of local varieties, known as landraces, which have great cultural and humanitarian value. Locally adapted landraces have sustained rural communities for centuries, and they are widely viewed as an important genetic resource for future breeding. Mexico is regarded as a center of diversity for maize germplasm, and many researchers have argued that on-farm conservation of this resource needs much greater emphasis and support than the Mexican government can provide (e.g., Bellon and Berthaud 2004). In addition, maize plays a central role in the country's heritage and culture, especially among the nation's numerous indigenous populations.

Because of many interacting scientific, socioeconomic, and political reasons, Mexico has banned the cultivation of genetically modified maize within its borders, at least for the present. Nonetheless, transgenes from U.S. maize were detected in isolated valleys near Oaxaca (Quist and Chapela 2001), raising questions about how the transgenes dispersed there, which specific transgenes were present, what effects they might have on local landraces and teosinte, and whether they could ever be

removed if this was deemed necessary. It is widely assumed that the transgenes found in Mexican landraces are derived from U.S. sources. Seeds
could be carried across the border, or they could arrive in the millions of
metric tons of U.S. maize that Mexico imports each year. Once an introduction occurs, the transgenes' ability to persist and multiply among landraces depends on whether transgenic plants have an inherent reproductive advantage and whether farmers actively select transgenic plants when
they save and trade seed.

The largest concern that has been voiced about the presence of transgenes in landraces of maize in Mexico is that this process could threaten
genetic diversity. This is a contentious issue, in part because terms like
genetic diversity and *genetic identity* have different meanings to different
people. If you believe that transgenes are inherently dangerous and represent genetic pollution, it is logical to conclude that the genetic composition of landraces is altered profoundly when transgenes are detected in
the plants' DNA. However, the scientific definition of *genetic diversity*
has a specific biological meaning that is directly relevant to crop breeding—it is the sum of all the variants of each gene in the gene pool of a
given population, variety, or species. This diversity represents tens of
thousands of genes, many of which vary within and among populations.
This is the genetic diversity that crop breeders and seed-saving farmers
need for future generations, so they can select for plants that can cope
with diseases, insect problems, competing weeds, and changing environmental conditions, while also maintaining the crop's nutritional and cultural value.

Contrary to what might be expected, the presence of a few novel
transgenes is not expected to have any direct negative effects on the overall amount of genetic diversity of landraces (Bellon and Berthaud 2004).
This conclusion derives from the biology of how genetic diversity is maintained in open-pollinated, farmer-selected landraces of maize. Locally
adapted varieties do not represent frozen-in-time heirloom stocks because rural Mexican farmers often trade seeds, plant mixtures of seeds
from different sources—including an occasional modern hybrid—and
allow cross-pollination to occur by planting different types of maize near
each other. Despite their intentional encouragement of gene flow, farmers are able to select and perpetuate recognizably different maize varieties
in much the same way that their ancestors did. If transgenes are added to
landraces, they are not expected to displace other genes. Instead, they

would be added to the mixture of genes that constitutes this dynamic, variable, and continually evolving genetic resource.

Even a highly favored transgene (i.e., one conferring characteristics that farmers tend to prefer) would not significantly displace other genes. This is because maize has high rates of outcrossing and genetic recombination. Initially, a new transgene would be linked to other genes on a chromosome originating from a modern maize variety. Then, sexual reproduction and recombination would act to break up these "linkage groups," such that countless new combinations of genes are present in each seed of a plant's progeny. Only the presence of large numbers of favored transgenes in many genomic locations would begin to drag along sizable regions of DNA from modern hybrids and thereby displace native genes.

Landraces could be threatened by other causes, such as being inundated with pollen from modern varieties growing around them (Ellstrand 2003). This scenario may be relatively uncommon in Mexico, at least so far, and if transgenic maize is deregulated, it may or may not have any exacerbating effect on the frequency of "gene swamping." Landraces also could disappear if farmers simply stop planting them, which, again, may or may not be influenced by the presence of genetically modified maize in Mexico (Bellon and Berthaud 2004).

This is not to say that scientists have all the answers about how transgenic crops will affect the genetic diversity of landraces and wild ancestors in the crop's center of origin—far from it, given the several sources of uncertainty that I have already mentioned. For example, farming practices that maintain landrace diversity depend on the active participation of rural farmers, whose behavior might be influenced by fear of transgene contamination or socioeconomic pressures that are indirectly related the increasing use of transgenics in modern agriculture (Bellon and Berthaud 2004; Gepts and Papa 2003). Socioeconomic factors clearly play an important role in how crop genes disperse and how much genetic diversity is maintained in farmers' fields. Another problem is that very little theoretical or empirical research has focused on the extent to which gene flow from crops, and transgenic crops in particular, affects the genetic diversity of recipient populations (e.g., Gepts and Papa 2003; Ellstrand 2003; Haygood, Ives, and Andow 2003). Clearly, these underfunded and underexplored areas of research need further attention from international and interdisciplinary teams of scientists.

A Research Agenda for Mexico

Before genetically modified maize is intentionally introduced into Mexico, each type of genetically modified maize should be evaluated carefully for potential health and environmental effects in the context of Mexican diets and the agricultural and natural landscapes where maize is cultivated. Research should also address questions about effects on nontarget species and the consequences of gene flow to teosinte. In fact, these types of studies are needed immediately because it is clear that some types of genetically modified maize have already entered Mexico through the back door. Scientific studies of the extent of transgene flow into maize and teosinte populations have been limited and controversial (see chapter 8), and this uncertain situation needs to be resolved quickly. It would be extremely helpful to know whether transgenes already have spread widely or whether they are present in very low frequencies, and which types of transgenes are present in Mexico.

Several Mexican farmer organizations, Greenpeace, and other environmental groups in Mexico have strenuously objected to the presence of transgenes in landraces of maize. The North American Free Trade Agreement has a mechanism for examining environmental consequences of free trade, and its Commission on Environmental Cooperation has examined the issue of genetically modified maize and biodiversity in Mexico, after being petitioned by these opponents of genetically modified crops. The findings and recommendations for the Environmental Cooperation Advisory Group have sparked further debate about genetically modified maize (available at http://www.cec.org/home). Although this controversy is not likely to be resolved any time soon, the more people can learn about the scientific, ethical, and political issues involved, the more likely it is that some level of understanding can be reached, a reasonable compromise achieved, and appropriate action taken.

Although gene flow from transgenic crops will continue to raise questions in a variety of contexts, many agree that a wide range of safe and beneficial uses of genetically modified crops can be put into practice worldwide. In fact, developing countries could benefit greatly from transgenic methods that enhance the yields and nutritional value of their staple crops. Greater international cooperation is needed to ensure that these crops are introduced in an ethical and environmentally sound manner, with thorough and rigorous evaluations of their potential risks and benefits.

References

Beckie, H., S. Warwick, H. Nair, G. Séguin-Swartz. 2003. Gene flow in commercial fields of herbicide-resistant canola (*Brassica napus*). *Ecological Applications* 13:1276–94.

Bellon, M. R., and J. Berthaud. 2004. Transgenic maize and the evolution of landrace diversity in Mexico: The importance of farmers' behavior. *Plant Physiology* 134:883–88.

Dalton, R. 2002. Superweed study falters as seed firms deny access to transgene. *Nature* 419:655.

Ellstrand, N. C. 2003. *Dangerous liaisons? When cultivated plants mate with their wild relatives.* Baltimore: Johns Hopkins University Press.

Gepts, P. 2002. A comparison between crop domestication, classical plant breedings, and genetic engineering. *Crop Science* 42:1780–90.

Gepts, P., and R. Papa. 2003. Possible effects of (trans)gene flow from crops on the genetic diversity of landraces and wild relatives. *Environmental Biosafety Research* 2:89–103.

Haslberger, G. A. 2003. Codex guidelines for GM foods include the analysis of unintended effects. *Nature Biotechnology* 21:739–41.

Haygood R., A. R. Ives, and D. A. Andow. 2003. Consequences of recurrent gene flow from crops to wild relatives. *Proceedings of the Royal Society of London, Series B* 270:1879–86.

ISB (Information Systems for Biotechnology). 2004. Databases of U.S. and international field tests of GMOs. http://www.isb.vt.edu/ (accessed June 4, 2004).

Klein, E. K. et al. 2003. Corn pollen dispersal: Quasi-mechanistic models and field experiments. *Ecological Monographs* 73:131–50.

Mellon, M., and J. Rissler. 2004. Gone to seed: Transgenic contaminants in the traditional seed supply. *Union of Concerned Scientists.* http://www.ucsusa.org/food_and_environment/biotechnology/page.cfm?pageID=1315 (accessed June 4, 2004).

NRC (National Research Council). 1987. *Field testing genetically modified organisms: Framework for decisions.* Washington, D.C.: National Academies Press.

———. 2000. *Genetically modified pest-protected plants: Science and regulation.* Washington, D.C.: National Academies Press.

———. 2002. Environmental effects of transgenic plants: The scope and adequacy of regulation. Washington, D.C.: National Academies Press.

———. 2004. *Safety of genetically engineered foods: Approaches to assessing unintended health effects.* Washington, D.C.: National Academics Press.

Partridge, M., and D. J. Murphy. 2004. Detection of genetically modified soya in a range of organic and health food products: Implications for the accurate labeling of foodstuffs derived from potential GM crops. *British Food Journal* 106:166–80.

Pilson, D., and H. R. Prendeville. In press. Ecological effects of transgenic crops and the escape of transgenes into wild populations. *Annual Review of Ecology and Systematics.*

Quist, D., and I. H. Chapela. 2001. Transgenic DNA introgressed into traditional maize landraces in Oaxaca, Mexico. *Nature* 414:541–43.

———. 2002. Biodiversity (Communications arising [reply]): Suspect evidence of transgenic contamination/maize transgene results in Mexico are artifacts. *Nature* 416 (April 11): 602.

Royal Society of Canada. 2001. Report of the expert panel on the future of food biotechnology. http://www.rsc.ca/foodbiotechnology/indexEN.html (accessed June 4, 2002).

Saxena, D., and G. Stotzky. 2001. *Bacillus thuringiensis* (Bt) toxin released from root exudates and biomass of Bt corn has no apparent effect on earthworms, nematodes, protozoa, bacteria, and fungi in soil. *Soil Biology and Biochemistry* 33:1225–30.

Snow, A. A. 2002. Transgenic crops: Why gene flow matters. *Nature Biotechnology* 20:542.

———. 2003. Genetic engineering: unnatural selection. *Nature* 424:619.

Snow, A. A. et al. 2003. A Bt transgene reduces herbivory and enhances fecundity in wild sunflowers. *Ecological Applications* 13:279–86.

———. 2004. Genetically engineered organisms and the environment: Current status and recommendations. Position paper of the Ecological Society of America. *Ecological Applications.* http://www.esa.org/pao/esaPositions/Papers/geo_position.htm (accessed June 4, 2004).

WHO (World Health Organization). 2003. Codex Alimentarius Commission. Joint FAO/World Health Organization Food Standard Program. Codex Ad Hoc Intergovernmental Task Force on Foods Derived from Biotechnology. http://www.who.int/foodsafety/biotech/consult/en/ (accessed June 4, 2004).

Wolfenbarger, L. L., and P. R. Phifer. 2000. The ecological risks and benefits of genetically modified plants. *Science* 290:2088–93.

7

Introduction of Transgenic Crops in Centers of Origin and Domestication

Paul Gepts

In 2003, farmers worldwide planted transgenic crops over about sixty-five million hectares, or 5 percent of total arable land area (James 2003; Food and Agriculture Organization 2004). Most transgenic crops are grown in four countries, the United States, Argentina, Canada, and China. The United States and Argentina together account for nearly 90 percent of transgenic production, with Canada and China accounting for most of the remainder. In the United States, the major transgenic crops are herbicide-tolerant soybean (*Glycine max*), lepidopteran insect–resistant cotton (mainly upland cotton, or *Gossypium hirsutum*), and lepidopteran insect–resistant and herbicide-tolerant maize (*Zea mays*); in Argentina, herbicide-tolerant soybean; in Canada, herbicide-tolerant canola; and in China, lepidopteran-resistant cotton. The shared characteristic of these countries is that none is actually located in the center of genetic diversity and domestication of their respective transgenic crops. Cotton and maize originated in Mesoamerica (Brubaker and Wendel 1994; Wilkes 2004), soybean in China (Shimamoto et al. 2000; Li et al. 2001), and oilseed rape presumably in Europe (Sauer 1993). With the exception of oilseed rape, which originated a few centuries ago (Sauer 1993) and has not been fully domesticated yet, the other crops have a history that stretches through several millennia in their respective centers of domestication.

Since Vavilov (1926),[1] it has been known that genetic diversity of crops is unequally distributed across the globe. For many crops, it is possible to identify certain areas with a high level of genetic diversity compared with other areas. Often, these areas correspond also to the center of domestication, namely, the process whereby a wild plant is subjected to a selection process conducted under human influence to increase adaptation to cultivated conditions and usefulness to consumers of the harvested products such as grains, fruits, and fibers. Domestication also includes selection for adaptation to new environments, as crops were dispersed from their original centers of domestication to other regions or continents (Gepts 2004a). Many of the domestication centers are actually located in megadiversity centers. Of the seventeen megadiverse countries, at least ten belong to a center of crop domestication (Brazil, China, Colombia, Ecuador, India, Indonesia, Malaysia, Mexico, Peru, and Venezuela).

There is no a priori reason why the introduction in a center of crop domestication of a new cultivar, even a transgenic one, should be cause for alarm. However, several aspects distinguish centers of domestication from other areas where a crop is grown. The different aspects include environmental, agricultural, sociocultural, and intellectual property rights issues. In this chapter, I will discuss each of these aspects and argue that the introduction of transgenic crops into centers of domestication should proceed only with caution, if at all. Many of my examples will address the situation of maize in its homeland (now called Mexico). However, similar arguments can be used for other crops in their respective centers of origin.

Environmental Issues

Gene Flow and Genetic Diversity

The foremost environmental issue is the presence of sexually cross-compatible relatives, whether domesticated or wild. The wild types may be directly related to a crop as progenitors or they may be indirectly related as neighboring taxa. Domesticated relatives are local, farmer-selected cultivars, also called landraces. Both wild and domesticated relatives fulfill important roles as reflections of sociocultural identities, production capital of farmers, and repositories of genetic diversity for plant breeders and farmers alike.

An important feature of these domesticated or wild relatives is that they generally cross readily with introduced cultivars. This feature sets the stage for potentially extensive gene flow in domestication centers between transgenic cultivars and their relatives. On the one hand, crops have evolved to increase self-pollination, which would reduce gene flow among crop varieties.[2] One the other hand, relatives of transgenic crops may have a more extended flowering time, thus increasing the probability of gene flow. In addition, for animal-pollinated crops, the presence of insect or other animal pollinators that have coevolved with the plant host in centers of domestication may also increase the potential for outcrossing.

Transgenic cultivars present certain issues that are unique and differ from nontransgenic cultivars in terms of the introduction of transgenes through gene flow. It has often been stated that the transformation process does not carry any inherent risks that do not exist in conventional, sexual transfer. Therefore, the product of the gene transfer, rather than the gene transfer process itself, should be regulated. Carrying this idea to a logical conclusion suggests that certain products of classical plant breeding should also be regulated (Gepts 2002). However, there is a dearth of information about the stability of insertion and expression of transgenes in new genetic backgrounds, especially in centers of domestication where genetic backgrounds may differ considerably from those in which transgenes were originally introduced. It also remains to be seen whether and to what extent this concern extends to nontransgenic cultivars as well.

Gene flow from transgene cultivars to native materials in centers of domestication has two potential consequences. First is a risk of accumulation of different transgenes in these native materials (called stacking), which may then serve as relays for the unwanted introduction of transgenes to other plant materials, destined for food or organic production. This could be particularly true for pharmaceutical or industrial compounds, which are highly undesirable in the food chain. However, no foolproof methods yet exist for keeping food and nonfood uses of crops separate. Even in the current seed production systems, transgenes are contaminating nontransgenic seed stocks at a low but measurable level (Friesen, Nelson, and Van Acker 2003; Mellon and Rissler 2004). The problem is even more marked in centers of domestication because the possibilities of physical isolation are more limited, given the presence of sexually compatible relatives. Accumulation of transgenes may also lead to untested combinations of these genes in the same plant.

Second, gene flow may affect the genetic diversity of the landraces and wild relatives in a number of situations. A genetically uniform source population (such as an improved or hybrid cultivar), high and recurrent levels of migration from the source to the recipient population (i.e., landraces), short distances (depending on the flowering biology of each crop), and/or a combination of these factors can lead to a potentially severe reduction in genetic diversity of the recipient populations and even genetic assimilation (defined as the displacement of the local diversity by the incoming diversity). Transgenic cultivars would not have a monopoly of displacement of genetic diversity. Actually, the development of uniform, elite cultivars by classical breeding has reduced genetic diversity.

The key factor is the degree of uniformity of the improved cultivars. In recent decades the trend has been toward concentration of breeding activities in both the public and private sector. For example, research centers such as the International Maize and Wheat Improvement Center, the International Rice Research Institute, and the International Center for Tropical Agriculture have bred cultivars with wide adaptation that presumably can be grown over broad areas. In the United States, the seed industry has witnessed two rounds of consolidation induced by the availability of molecular biology tools and the application of intellectual property rights to living organisms and basic biological processes. Before this situation existed, breeding programs tended to be smaller and with a more local focus, which maintained a broader range of genetic diversity.

Ecosystem Effects

In addition to concerns about gene flow, it is important to consider that the environments in centers of domestication are quite different from those where transgenic cultivars are grown today, as illustrated by a brief discussion of Bt crops (transformed with the gene for the *Bacillus thuringiensis* [Bt] toxin). Not only are the pests in centers of domestication, like Mexico, different, but nontarget organisms (e.g., nonpest lepidopteran and coleopteran species in the case of the Bt toxin) are also quite different. Studies have primarily been conducted in the United States and Europe. For example, following the initial observation by Losey, Rayor, and Carter (1999) of the susceptibility of the monarch butterfly to the Bt toxin, more detailed analyses were conducted (after the regulatory release of transgenic maize), which concluded that the effects of Bt on the monarch butterfly were minimal in the short term in the conditions of the Midwest (Sears et al. 2001 and references therein) but not necessarily in the long term (Scriber 2001). Similarly detailed studies are lacking in domestication centers, so we do not know the implications for insects in these areas.

Letourneau, Hagen, and Robinson (2002) established a list of about 370 lepidopteran species associated with maize. Of these, only eleven had been examined for their susceptibility to the Bt toxin. Letourneau, Robinson, and Hagen (2003) evaluated what might happen if transgenes escaped into relatives of cotton, rapeseed, and rice by examining lists of sexually compatible relatives, host ranges of lepidopterous insects, their susceptibility to the Bt toxin, and information about the ability of these insects to limit plant growth. They concluded that data are insufficient to establish a risk of ecological release associated with the escape of transgenes among relatives of the three transgenic crops studied. An additional concern may be the effect on certain pollinating, parasitoid, and predator insects (Groot and Dicke 2002). Wolfenbarger and Phifer (2000) present a comprehensive view of the need to measure ecosystem risk and benefits resulting from the introduction of transgenic crops.

Agricultural Issues

Farmers in industrial and in traditional or subsistence agriculture (characteristic of the majority of farmers in centers of domestication) play different roles. In industrial agriculture, farmers have a more specialized role, limited to the production of crops. In contrast, in traditional agriculture, farmers play a role in conservation, development of new cultivars, and processing and consumption of crops in addition to crop production. Specifically, farmers in traditional agriculture play an active role in maintaining crop landraces (in situ conservation; Maxted, Ford-Lloyd, and Hawkes 1997). Landraces are defined as locally distributed and adapted domesticated plants, maintained by farmers. Farmers exert selection to maintain different types according to their use in different cropping systems and for different consumer uses.

Farmers are also willing to experiment by bringing in new materials, including improved bred varieties (e.g., Quiros et al. 1992; Bellon and Berthaud 2004), which then may cross with the local materials and generate new materials (process of *acriollamiento,* or creolization; Bellon and Risopoulos 2001). Farmers exchange seeds with each other, primarily with relatives but also with others in the same or neighboring villages or regions (e.g., Almekinders, Louwaars, and De Bruijn 1994; vom Brocke et al. 2003; Nkongolo 2003). Through their experimentation and selection, farmers may assure better adaptation of the planting materials to the local agroecological niches. Thus seeds are not merely an agricultural ingredient (like, for example, fertilizer or irrigation water); they are more aptly considered part of the agricultural capital of the farmer, just as land and equipment are.

Genetic diversity is a prerequisite for the development of superior cultivars by farmers and breeders alike. In addition, for farmers genetic diversity is also insurance against the vagaries of production conditions. Typically, a farmer may plant a mixture of cultivars that have different maturities and adaptations to assure some level of production. However, the continued existence of on-farm diversity is threatened by the loss of farmers through migration to cities and other countries, the spread of industrial monoculture cropping systems, and gene flow from or replacement by modern cultivars. Transgenic crops, to the extent that they are an inherent part of industrial agricultural systems, can be a driver in the potential reduction of genetic diversity. Recurrent gene flow from a uniform crop is more likely to displace native genetic diversity, as I mentioned earlier. The combination of intellectual property rights and molecular biology tools has made the development of transgenic cultivars by the private sector possible (Gepts 2004b). Concurrently, the seed industry has been consolidating, so that a few companies now dominate the seed market for several crops, such as maize and cotton. This market concentration raises the possibility that the elite domesticated gene pool will become even more depleted of genetic diversity (Gepts and Papa 2003).

To avoid such a situation in centers of domestication, the transgenic construct could be made available to breeders who could incorporate it into local varieties and thus maintain a more diverse genetic background. Such a situation exists for transgenic herbicide-resistant soybean in the United States. The glyphosate resistance transgene (used in Roundup Ready crop plants) was made available to many companies and public institutions. As a result, a large set of superior cultivars, representing the current diversity in nontransgenic North American soybean, was used to develop current herbicide-resistant cultivars (Sneller 2003). Individual breeding programs, whether public or private, run the risk of having a narrow genetic base. However, public lines originate from many independent programs and as a whole tend to be more representative of the entire range of elite genetic diversity. In addition, current public programs increase diversity through the use of exotic germplasm (Sneller 2003). Exotic germplasm are plant materials that come from different countries or continents. As such, they are generally not adapted to U.S. conditions, but they carry useful traits such as disease or pest resistance. The long-term focus of public breeding programs allows them to use exotic germplasm to introduce these useful traits into advanced cultivars. Exchanges among breeding programs are, therefore, essential to maintaining a gene pool of elite cultivars that is as broad-based as possible.

Minimizing reductions in genetic diversity in centers of domestication because of the use of advanced cultivars, whether transgenic or not, would require a diverse group of breeding programs that actively interchange breeding lines. Plant breeding has proved to be a very successful approach that has not lost any of its power. Adoption of modern breeding methods, such as marker-assisted selection, has greatly increased the power of traditional breeding methods. For example, in the common bean, several diseases such as white mold, golden mosaic virus, and common bacteria, once considered very difficult, if not intractable, are now amenable to genetic improvement through the use of marker-assisted selection, a broader range of germplasm, and improved screening methods (Urrea et al. 1996; Miklas et al. 2001; Kelly et al. 2003). In this respect, transgenic cultivars can make a contribution when screening has shown the native diversity to be insufficient and breeding to improve a critical trait has not worked. An example is the lack of resistance in maize to the European corn borer in the Midwest for which Bt maize provides a solution (Gepts 2002).

However, adoption of improved cultivars may be limited in certain locations. In Mexico, for example, 80 percent of the maize land is still planted with landraces rather than improved cultivars. There are several valid reasons for this limited adoption, including the varied topography (and attendant multitude of microniches), the underfunding of public breeding programs and agricultural research in general, and consumer preference, which is directed to very specific traits, such as colors, textures, cookability, and shelf-life. One solution to this problem might be to decentralize breeding programs to rural areas where farmers themselves would become more involved in the improvement of their local landraces in collaboration with plant breeders (a process also known as participatory plant breeding; Cleveland and Soleri 2002). Such an approach to plant breeding should be part of a broader goal of achieving self-sufficiency in maize production. In the case of maize in Mexico, A. Turrent (personal communication) has shown that it is possible to raise yield and total production to the point that Mexico becomes self-sufficient for its basic food crop (as well as for its nutritional complement, the common bean).

To make transgenes available—if and when necessary—to a broad section of these programs may require market segmentation for intellectual property rights. This means private companies would have to forgo their royalties for applications in developing countries in order to benefit smallholder farmers, as has been proposed for golden rice (Wai 2003). Another possibility is the public development of transgenic cultivars. The research agenda for transgenic crops should not be determined exclusively by the private sector in industrialized countries. Because the private sector primarily addresses crops with a large market and farmers who can afford to buy seeds, it may not address crops with a smaller market in countries with subsistence farmers. To put transgenic technologies fully to the test, these ought to be designed to fit the agronomic and socioeconomic conditions of smallholder farmers (Chrispeels 2000).

Intellectual Property Rights Issues

One factor driving the development of a transgenic seed industry in the United States and other industrialized countries has been the availability of intellectual property rights over living organisms (Gepts 2004b). The landmark Supreme Court decision in this area was *Diamond v. Chakrabarty* (447 U.S. 303 [1980]), which set the stage for the award of utility patents for crop cultivars.

The United States is one of only three countries (with Australia and Japan) to award utility patents for crop cultivars. Other countries provide only plant variety protection (PVP) certificates. Utility patents must fit the criteria of novelty, inventiveness, and utility. Unlike the PVP certificates, utility patents do not allow for research exemptions or farmer's exemptions. Research exemptions, as allowed by PVP certificates, allow researchers to use patented cultivars as parents in crosses to develop the next generation of improved cultivars. With a farmer's exemption, farmers could harvest patented seeds and replant them on their own land (a practice called seed saving), although they could not sell or give them away to others. Since these exemptions are not allowed by utility patents, developers of genetically modified seeds in the United States have increasingly patented those crop cultivars rather than obtaining the more flexible PVP certificates.

Although utility patents are not available for crop cultivars in most countries, transgenic constructs or methods broadly applicable to plants (i.e., not limited to a specific genotype) are patentable subject matter not only in the United States but in many other countries as well. For example, a transgenic construct carrying the Bt gene or a herbicide resistance gene can be patented. Although a more complete description of intellectual property rights on biodiversity is beyond the scope of this chapter, I do want to note that patent and plant variety protection rights are granted for a limited period (generally twenty years) and a specific place (they are limited to the country that awards them). This being said, patent rights are extremely strong—the courts generally frown on anything that might weaken these rights. For example, patent rights supersede property rights. Ignorance about a patent and lack of intent cannot be used as defense against an infringement accusation. Most surprisingly, gene flow cannot be used as a defense against infringement. Thus, if a company releases a transgenic cultivar, it is not now responsible for the inadvertent escape of transgenes to nontransgenic fields. However, a farmer can be held liable for patent infringement if the patented transgene inadvertently lands on his or her property. This has potential legal implications, especially in centers of domestication where gene flow is particularly widespread.

Although intellectual property rights are limited territorially, their existence nevertheless creates a series of challenges. First, industrialized countries have pushed less-developed countries (where most centers of domestication are located) to adopt intellectual property rights legislation through such mechanisms as the Trade-Related Intellectual Property (TRIPS) agreement, administered by the World Trade Organization (WTO). By joining the WTO, a country commits to the development and enforcement of intellectual property rights legislation. Specifically with regard to crop cultivars, the TRIPS agreement requires countries to provide protection for these cultivars, although not necessarily patenting. The system most often proposed is similar to plant variety protection. Transgenic constructs are still subject to patenting.

The stipulations of intellectual property rights for crop cultivars are in direct conflict with practices of many farmers in centers of domestication. In traditional agriculture, seed stocks are readily exchanged and are a public good shared by individuals in communities. This contrasts with individual inventorship and assignment to companies or institutions in industrialized countries. Landraces have been handed down as heirlooms for generations (Zimmerer 1996; Louette and Smale 2000; Perales,

Brush, and Qualset 2003), a practice that also makes identification of individual inventors difficult, if not impossible. Furthermore, many landraces are actually mixtures of genotypes and not pure lines, which would therefore not fit the criteria for plant variety protection. Among the standard practices of farmers are to exchange seed materials and let cross-pollination recombine different genotypes, not only in cross-pollinated species but also in self-pollinated species such as the common bean (Bellon and Risopoulos 2001; Perales, Brush, and Qualset 2003; D. Zizumbo and P. Colunga GarcíaMarín, personal communication). In other words, in traditional agriculture, gene flow is a widely accepted feature or practice, whereas in industrialized agriculture it is to be avoided in order to avoid legal troubles related to intellectual property rights or contamination of the seed stock or grains. Thus introduction of Western intellectual property rights legislation in developing countries creates the possibility that local or indigenous farmers in centers of domestication could be subjected to legal action by the patent holder.

A further consideration is traditional knowledge associated with landraces. Traditional knowledge refers to information held by local or indigenous people, in this case with regard to biodiversity (Brush and Stabinsky 1996). Traditional knowledge is an inherent part of biodiversity and a resource in its own right. For example, Fabricant and Farnsworth (2001) determined that 80 percent of plant-based drugs in Western medicine have had an ethnomedical (i.e., non-Western) use identical or related to the current use of the active elements of the plant. With regard to crops, traditional knowledge encompasses information about their agronomic or culinary characteristics. Traditional knowledge is an essential aspect of an indigenous group's cultural survival; it has been developed through generations of intimate contact with the biological materials (Mauro and Hardison 2000). Traditional knowledge is not, however, limited to the knowledge of indigenous people but encompasses knowledge (and associated heirloom varieties) of local, nonindigenous communities in modern societies as well (e.g., Bérard and Marchenay 1996).

Thus indigenous societies or local farmer groups often practice an informal system of innovation and information dissemination, which does not fit well into a Western-style intellectual property rights system, nor does the latter offer rewards for past efforts in innovation and conservation that serve as a foundation for the existence of biodiversity in general and crop biodiversity in centers of diversity in particular. The distinct features of the use and conservation of biodiversity in developing countries have led to a call for a separate legal system that recognizes the contributions of indigenous or local communities. When dealing with crop landraces, this legal system refers to farmers' rights. However, little progress has been made in developing an enforceable legal framework to support farmers' rights in practice (Gepts 2004b).

Cultural Issues

The long and intimate coexistence of people and crops in centers of domestication is reflected in an extensive cultural presence of the crops among the people, indigenous or not, living in these centers. Maize, for example, has multiple food uses in Mexico, its center of domestication (e.g., Kennedy 2003). Its husks are used as wrapping for dishes, its stalks and leaves as forage, and so on. Mexican Spanish contains an abundance of words derived from pre-Hispanic languages. These words are closely related to the preparation and consumption of maize and attest to the long cultural history of the crop in its center of domestication (Salvador 1997). The importance of basic food crops in their center of domestication is reflected also in their inclusion in creation beliefs. The Popol Vuh (Tedlock 1996), the creation story of the Quiché Maya, relates how, after several failed attempts based on different starting materials, Heart of Sky successfully made humans out of maize. Similar observations can be made for other crops in their respective centers of domestication, such as wheat in southwestern Asia and rice in eastern Asia.

This long-term, close association between people and their respective crops in centers of domestication explains some of their behavior, which at first may seem incomprehensible to outsiders. For example, cultivation of maize in Mexico sometimes takes place despite the lack of economic incentives and returns (Perales, Brush, and Qualset 2003). Rather, noneconomic motives such as consumer preferences (color, flavor, cooking quality, shelf life before and after cooking) and cultural identity play an important part as well. Breeding programs, whether they involve transgenic techniques or not, should take these preferences into account. It is not sufficient to consider productivity alone (yield potential, tolerance to abiotic stresses, resistance to biotic stresses). In addition, continued cultivation of maize, a major food crop, can be justified as insurance in the face of uncertain market conditions, which are characterized by uncertain employment and fluctuating prices, induced in part by international trade agreements.

Emphasis on qualities appreciated by the consumer, in addition to those of importance to the producer, may also be a strategy to assure both the conservation of genetic resources and revenues to the farmer. The European Union has, for example, developed specific designations, such "protected geographic indication" or "protected designation of origin," which could protect local genetic resources and make their product better known. About five hundred cheese, meat, fruit, and vegetable products are registered as protected geographic indications or protected designations of origin. It remains to be determined whether such attempts at maintaining agricultural and culinary traditions are compatible with the use of transgenic cultivars.

Human and Animal Health Issues

It is beyond the scope of this chapter to address issues related to human and animal health. However, several arguments suggest that these issues need to be addressed in the context of centers of domestication. For example, the genetic composition of human consumers, and therefore the intrinsic reactions to different components included in foodstuffs, may differ from those existing in developed countries such the United States, where transgenic cultivars have been tested initially. Because some crops are staple crops in centers of domestication, the exposure may vary from that experienced by human populations in the United States or other countries.

Conclusions

Several issues, including environmental, agronomic, and intellectual property rights, suggest that the introduction of transgenic crops in their respective centers of domestication requires specific attention beyond that devoted to these crops outside the centers of domestication.

A dearth of experimental data often hampers the evaluation of potential risks associated with the introduction of transgenic crops in centers of diversity. Such studies need to be conducted before the introduction of transgenes in domestication centers.

Given several issues that have been raised here, those who want to introduce transgenic cultivars into a center of genetic diversity and domestication ought to be required to prove that they are safe and can be controlled. There may well be cases in which other approaches, whether genetic or not, will solve the problem while circumventing the issues raised by transgenic cultivars. In turn, these other approaches should also be subjected to comparative risk-benefit analyses.

Delaying or denying the introduction of transgenic crops in centers of origin does not amount to denying the benefits of genetic improvement to the people of these centers. In most cases, classical plant breeding provides a functional alternative that has stood the test of time, although in some limited cases its environmental and human health effects may also need to be monitored.

Transgenic cultivars could play a role if they are specifically designed to address constraints faced by smallholder farmers and fit into the agronomic, environmental, public health, and consumer preferences characteristic of their centers of domestication.

Acknowledgments

I thank Matthew Hufford, Daniel Kleinman, and Abby Kinchy for useful comments. The views expressed in this contribution are mine and not those of my institution or funding agencies.

Notes

1. Nikolai Vavilov (1887–1943), the former director of the All-Union Institute of Plant Industry in St. Petersburg, Russia, was a prominent Russian crop geographer who led countless explorations in Eurasia, Africa, and the Americas. Based on these explorations, he formulated, among others, the theory of the centers of origin of cultivated plants (1926).

2. Plants are characterized by three major reproductive systems. In selfing species (also known as autogamous, or self-pollinating, species), the pollen of a flower is involved in fertilization of the ovules of the same flower. In outcrossing species (also known as allogamous species), pollen is transferred to flowers of other individuals, generally by wind or animals (such as insects or birds). One should keep in mind that the reproductive system of plants may vary to a certain degree. For example, selfing species will generally exhibit some degree of outcrossing, and vice versa for outcrossing species. The transfer of genes, either by pollen or seed, to a new population or location is called gene flow.

References

Almekinders, C. J. M., N. P. Louwaars, and G. H. De Bruijn. 1994. Local seed systems and their importance for an improved seed supply in developing countries. *Euphytica* 78:207–16.

Bellon, M. R., and J. Berthaud. 2004. Transgenic maize and the evolution of landrace diversity in Mexico: The importance of farmers' behavior. *Plant Physiology* 134:883–88.

Bellon, M. R., and J. Risopoulos. 2001. Small-scale farmers expand the benefits of improved maize germplasm: A case study from Chiapas, Mexico. *World Development* 29:799–811.

Bérard, L., and P. Marchenay. 1996. Tradition, regulation and intellectual property: Local agricultural products and foodstuffs in France, pp. 230–43. In *Valuing local knowledge,* edited by S. B. Brush and D. Stabinsky.

Brubaker, C., and J. Wendel. 1994. Reevaluating the origin of domesticated cotton (*Gossypium hirsutum,* Malvaceae) using nuclear restriction fragment-length polymorphisms (RFLPs). *American Journal of Botany* 81:1309–26.

Brush, S. B., and D. Stabinsky (eds.). 1996. *Valuing local knowledge: Indigenous people and intellectual property rights.* Washington, D.C.: Island Press.

Chrispeels, M. 2000. Biotechnology and the poor. *Plant Physiology* 124:3–6.

Cleveland, D., and D. Soleri (eds.). 2002. *Farmers, scientists, and plant breeding: Integrating knowledge and practice.* Wallingford, England: CABI.

Fabricant, D. S., and N. R. Farnsworth. 2001. The value of plants used in traditional medicine for drug discovery. *Environmental Health Perspectives* 109 (Suppl 1): 69–75.

Food and Agriculture Organization. 2004. Statistical databases. *Food and Agricultural Organization.* http://faostat.fao.org/faostat/collections?subset=agriculture (accessed March 17, 2004).

Friesen, L. F., A. G. Nelson, and R. C. Van Acker. 2003. Evidence of contamination of pedigreed canola (*Brassica napus*) seedlots in western Canada with genetically engineered herbicide resistance traits. *Agronomy Journal* 95:1342–47.

Gepts, Paul. 2002. A comparison between crop domestication, classical plant breeding, and genetic engineering. *Crop Science* 42:1780–90.

———. 2004a. Domestication as a long-term selection experiment. *Plant Breeding Reviews* 24 (Part 2): 1–44.

———. 2004b. Who owns biodiversity and how should the owners be compensated? *Plant Physiology.* 134:1295–1307

Gepts, Paul, and R. Papa. 2003. Possible effects of (trans)gene flow from crops on the genetic diversity from landraces and wild relatives. *Environmental Biosafety Research* 2:89–103.

Groot, A. T., and M. Dicke. 2002. Insect-resistant transgenic plants in a multitrophic context. *Plant Journal* 31:387–406.

James, C. 2003. Preview: Global Status of Commercialized Transgenic Crops: 2003. *International Service for the Acquisition of Agri-biotech Applications.* ISAAA Briefs No. 30. http://www.isaaa.org/kc/CBTNews/press_release/briefs30/es_b30.pdf (accessed March 14, 2004).

Kelly, J. D., P. Gepts, P. N. Miklas, and D. P. Coyne. 2003. Tagging and mapping of genes and QTL and molecular marker-assisted selection for traits of economic importance in bean and cowpea. *Field Crops Research* 82:135–54.

Kennedy, D. 2003. *From my Mexican kitchen: Techniques and ingredients.* New York: Clarkson Potter.

Letourneau, D. K., J. Hagen, and G. S. Robinson. 2002. Bt-crops: evaluating the benefits under cultivation and risks from escaped transgenes in the wild, pp. 33–98. In *Genetically engineered organisms: assessing environmental and human health effects,* edited by D. K. Letourneau and B. E. Burrows. Boca Raton, Fla.: CRC Press.

Letourneau, D. K., G. S. Robinson, and J. Hagen. 2003. Bt crops: Predicting effects of escaped transgenes on the fitness of wild plants and their herbivores. *Environmental Biosafety Research* 2:219–46.

Li, Z. L. et al. 2001. Molecular genetic analysis of U.S. and Chinese soybean ancestral lines. *Crop Science* 41:1330–36.

Losey, J., L. Rayor, and M. Carter. 1999. Transgenic pollen harms monarch larvae. *Nature* 399: 214.

Louette, D., and M. Smale. 2000. Farmers' seed selection practices and traditional maize varieties in Cuzalapa, Mexico. *Euphytica* 113:25–41.

Mauro, F., and P. D. Hardison. 2000. Traditional knowledge of indigenous and local communities: International debate and policy initiatives. *Ecological Applications* 10: 1263–69.

Maxted, N., B. Ford-Lloyd, and J. Hawkes. 1997. *Plant genetic conservation: The in situ approach.* London: Chapman and Hall.

Mellon, M., and J. Rissler. 2004. *Gone to seed.* Cambridge, Mass.: Union of Concerned Scientists.

Miklas, P., W. Johnson, R. Delorme, and P. Gepts. 2001. QTL conditioning physiological resistance and avoidance to white mold in dry bean. *Crop Science* 41:309–15.

Nkongolo, K. K. 2003. Genetic characterization of Malawian cowpea (*Vigna unguiculata* [L.] Walp) landraces: Diversity and gene flow among accessions. *Euphytica* 129:219–28.

Perales, H., S. B. Brush, and C. O. Qualset. 2003. Dynamic management of maize landraces in central Mexico. *Economic Botany* 57:21–34.

Quiros, C. F. et al. 1992. Increase of potato genetic resources in their center of diversity: The role of natural outcrossing and selection by the Andean farmer. *Genetic Resources & Crop Evolution* 39:107–13.

Salvador, R. J. 1997. Maize. *The Maize Page.* http://maize.agron.iastate.edu/maizearticle.html (accessed March 17, 2004).

Sauer, J. 1993. *Historical geography of crop plants.* Boca Raton, Fla.: CRC Press.

Scriber, J. M. 2001. Bt or not Bt: Is that the question? *Proceedings of the National Academy of Sciences* 98:12328–30.

Sears, M. K. et al. 2001. Impact of Bt corn pollen on monarch butterfly populations: A risk assessment. *Proceedings of the National Academy of Sciences* 98:11937–942.

Shimamoto, Y. et al. 2000. Characterizing the cytoplasmic diversity and phyletic relationship of Chinese landraces of soybean, Glycine max, based on RFLPs of chloroplast and mitochondrial DNA. *Genetic Resources & Crop Evolution* 47: 611–17.

Sneller, C. H. 2003. Impact of transgenic genotypes and subdivision on diversity within elite North American soybean germplasm. *Crop Science* 43:409–14.

Tedlock, D. 1996. *Popol Vuh: The definitive edition of the Mayan book of the dawn of life and the glories of gods and kings.* New York: Touchstone.

Urrea, C., P. Miklas, J. Beaver, and R. Riley. 1996. A codominant randomly amplified polymorphic DNA (RAPD) marker useful for indirect selection of bean golden mosaic virus resistance in common bean. *Journal of the American Society for Horticultural Science* 121:1035–39.

Vavilov, N. I. 1926/1992. Centers of origin of cultivated plants, pp. 22–135. In *Origin and geography of cultivated plants,* edited by V. F. Dorofeyev and translated by D. Löve. Cambridge: Cambridge University Press.

vom Brocke, K. et al. 2003. Farmers' seed systems and management practices determine pearl millet genetic diversity patterns in semiarid regions of India. *Crop Science* 43:1680–89.

Wai, T. 2003. IRRI: The experience of an international public research institute. *International Union for the Protection of New Varieties of Plants.* http://www.upov.int/en/documents/Symposium2003/wipo_upov_sym_14.pdf (accessed March 17, 2004).

Wilkes, G. 2004. Corn, strange and marvelous: But is a definitive origin known? pp. 3–63. In *Corn: Origin, history, technology, and production,* edited by C. Smith. New York: John Wiley.

Wolfenbarger, L. L., and P. R. Phifer. 2000. The ecological risks and benefits of genetically engineered plants. *Science* 290:2088–93.

Zimmerer, K. 1996. *Changing fortunes: Biodiversity and peasant livelihood in the Peruvian Andes.* Berkeley: University of California Press.

8

Agricultural Biotechnology Science Compromised

The Case of Quist and Chapela

Kenneth A. Worthy, Richard C. Strohman,
Paul R. Billings, and the Berkeley
Biotechnology Working Group

In November 2001, biologists David Quist and Ignacio H. Chapela (2001) generated international controversy when they published research findings demonstrating the presence of transgenes (genes incorporated into the genome of an organism through genetic engineering) in maize landraces in Mexico. Their results were both unexpected and important for a number of reasons. There had been a moratorium on planting of transgenic maize in Mexico since 1998, there were no known pathways for transgenes to enter maize landrace genomes in the field, and the landraces have critical material and symbolic importance. Following their initial report (in the journal *Nature*), there were condemnations of their results and methods, public accusations of incompetence, the publication in *Nature* of two particularly harsh and dismissive critiques accompanied by an unprecedented pseudo-retraction of the original report in an "Editorial Note," a joint statement by many scientists regarding research quality, and so on. In this chapter, we examine the controversy surrounding Quist and Chapela's report and show that this controversy exposed systemic problems in the science of agricultural biotechnology. We additionally show that their research was valuable both in correctly revealing the contamination of maize landraces by transgenes and in opening up a discussion about the behavior of transgenes in ecological contexts.[1]

Maize landraces are cultivars of maize that have evolved with and have been genetically improved by traditional agriculturalists in central Mexico for millennia, without the intervention of modern breeding techniques. Quist and Chapela's paper raised potentially grave concerns for several reasons. First, central Mexico is the world's center of genetic diversity for maize, supplying genetic resources that are vital to the continued development and persistence of maize cultivars throughout the world. Second, no one yet knows how transgenes will affect maize landraces; some are concerned that the introduction of transgenes into landraces, such as those that convey survival advantages to some cultivars, could adversely affect genetic diversity (Rissler and Mellon 1996, 112; Gepts, this volume). Finally, maize has profound material and spiritual meaning for Mexican culture. In a scientific and economic context in which agricultural biotechnology (ag-biotech) corporations and researchers seek to expand the use of transgenic organisms around the globe and are encountering increasing political and cultural opposition, Quist and Chapela's research came under intense scrutiny.

As occurs normally in scientific discourse, some errors in Quist and Chapela's research that went undetected in the initial peer-review process have subsequently surfaced; indeed, Quist and Chapela acknowledge several problems in their report (Quist and Chapela 2002). Yet these flaws alone cannot explain the intensity of negative reactions, which extended well beyond those published in *Nature*. Quist and Chapela suffered public accusations of basic laboratory incompetence in major newspapers and elsewhere (Yoon 2002). The Bivings Group, a public relations firm with ties to Monsanto, apparently made a "viral marketing" effort in which they created false Internet identities and spread rumors to prime the assault on Chapela on AgBioWorld, a Listserv used by more than three thousand scientists worldwide. In that assault, otherwise unknown individuals with e-mail addresses that were traced to the Bivings Group posted messages containing unsupported claims including that the *Nature* paper had not been peer reviewed and that Chapela was "first and foremost an activist," also suggesting that he was well paid for helping with "misleading fear-based marketing campaigns." (Monbiot 2002).

Regrettably, much of this opposition has detracted from the healthy scientific process of improving the quality of data, sharpening analysis, and conducting further research to answer emerging questions. We are confronted instead by an atmosphere of hostility and mistrust that upstages a critical research question: When transgenic DNA moves unintentionally into new species or new environments, what are the evolutionary, ecological, genetic, and social consequences? This chapter reveals many of the political and economic conflicts surrounding the controversy that compromise the integrity of the science surrounding ag-biotech.

To shed light on the controversy, it is helpful to review first the scientific claims made by Quist and Chapela and their respondents. The first finding of Quist and Chapela—the presence of transgenic DNA constructs in Mexican maize landraces—stands unrefuted. It was reported as confirmed by tests run by Mexican government researchers that even produced estimated percentages of contamination (Brown 2002; Enciso L. 2002) and was confirmed as well by Quist and Chapela's DNA hybridization test, which addressed the only substantive criticism—that of contamination of research materials in the process of conducting a polymerase chain reaction (PCR; Quist and Chapela 2002).[2] In light of Quist and Chapela's presentation of additional data, even the ag-biotech promotional organization AgBioWorld, which organized opposition to Quist and Chapela as demonstrated by the group's "Joint Statement," agrees that this first finding stands undisputed (AgBioWorld 2002a, 2002b). None of the criticisms of Quist and Chapela's techniques published in *Nature* effectively refutes the first finding of transgenes in the landraces. This may not be apparent to the casual reader, because the critiques highlight flaws rather than identifying how these flaws might have affected the overall results (Kaplinsky et al. 2002; Metz and Fütterer 2002a). Comments from a number of scientists have defused the impact of this finding by calling it "obvious" (Shouse 2002) and "inevitable and welcome" while simultaneously challenging "the methodology and results reported in the *Nature* paper" (see AgBioWorld 2002b for examples of both). These scientists attempt to have it both ways: The presence of transgenes in the environment is certain (thus Quist and Chapela's primary finding is trivial), yet it is also doubtful (thus Quist and Chapela are to be disregarded as scientists, even if their primary finding is correct).

Quist and Chapela do not attempt to address the question of how transgenes could have become introduced into maize landraces in central Mexico. That is a much larger issue that goes beyond the science involved

in determining the presence of transgenes and the scientific consequences of such. Mexico has prohibited all nonexperimental planting of
transgenic maize since 1998. Yet it has not prohibited the sale of transgenic maize for consumption—the United States has exported large
quantities of transgenic maize to Mexico as food. It may be that some of
this maize was planted by commercial or subsistence farmers in Mexico
and then became genetically incorporated into maize landraces through
pollination. Such introduction could have been deliberate for the purpose
of experimentation or accidental due to contamination of seed sources.
No definitive studies have been conducted to conclusively demonstrate
how transgenic maize became incorporated into Mexican maize landraces. Indeed, because of the pervasiveness of transgenic maize varieties
throughout the United States and several other countries, together with
the apparent inability of the producers of transgenic maize to prevent un-
planned movement of transgenes (through pollen drift, intentional planting, commingling, mislabeling, or the global transportation of grains), the
current and historical routes of transgene movement may never be adequately understood.

In the second section of their report, Quist and Chapela make an exploratory attempt to determine, on a molecular level, how transgenes have
become incorporated into maize landraces. They suggest that transgene
fragmentation occurred and that transgenic sequences were found in
multiple locations in the landrace genome. This would mean that not only
are transgenes spreading into non-target plants, but they are also establishing there in ways that are not well understood or controlled. From this
emerges an image of agricultural genetic engineering as a process that is
much less precise and controlled than some proponents claim. The implications of transgenic fragmentation have not been fully explored, but it
could conceivably lead to the accidental creation of novel genes.

While Quist and Chapela's work in this section is problematic—which
they acknowledged in their subsequent response to critics in Nature
(Quist and Chapela 2002)—several remarkable things stand out. First,
many of Quist and Chapela's critics have exaggerated the claims that the
authors made in the second section in order to discredit them—the common strawman rhetorical technique. Second, many of Quist and Chapela's
critics, including those published in Nature, have used the problematic
nature of Quist and Chapela's work in the second section to obscure the
more clearly supported finding of the first section—that transgenes have

become incorporated into maize landraces. Yet the finding of the first section does not depend on efforts outlined in the second section. Moreover, upon close examination, one result presented in the second section is valid and further reinforces the finding of the first section—a transgene was recovered again. Third, as problematic as the second section was, it did produce at least the one valid result in regard to transgene fragmentation. Yet critics did not acknowledge this result, nor did they use it as an occasion to call for further investigation.

In the problematic second section, Quist and Chapela report on experiments in which they use the inverse polymerase chain reaction (i-PCR) process to probe for the cauliflower mosaic virus 35S promoter (hereafter, CaMV promoter) and to sequence it along with adjacent DNA. (The CaMV promoter is used in the vast majority of transgenic crops because it enhances the expression of transgenes inserted downstream of it.) They used this method to investigate first whether transgenes were complete, changed, or fragmented and, second, where transgenes had been incorporated into the maize landrace genome. (Fragmentation and rearrangement of transgenes have been known to occur during insertion. It is not necessarily evidence of postinsertion rearrangement.) Using this method to determine location does not always produce definitive results, because sequences arising from these experiments must be compared to a database of historical genetic sequences. That database is incomplete; moreover, matches between the database and experimental samples are generally inexact because of genetic variations between individual organisms—in this case, individual maize plants.

After factoring in valid critiques, Quist and Chapela's data from these i-PCR experiments suggest that transgene fragmentation occurred, but where it happened is not clear. In their original report, Quist and Chapela's language clearly reflects the uncertainties associated with their tentative findings of fragmentation and multiple positioning: "*suggesting* that the promoter was inserted into the criollo [maize landrace] genome at multiple loci . . . *suggests* the occurrence of multiple introgression events, *probably* mediated by pollination. . . . The *apparent* predominance of reassorted sequences obtained in our study *might* be due to PCR bias for amplification of short fragments" (Quist and Chapela 2001, 542; emphasis added). This language contrasts with that of their primary finding of transgenic incorporation: "Our results *demonstrate* that there is a high level of gene flow from industrially produced maize towards populations of progenitor landraces" (Quist and Chapela 2001, 542; emphasis added). In their attacks, Kaplinsky, Metz, and their associates, along with other published critics of the study, falsely imply that Quist and Chapela made far stronger claims than they in fact did in the tentative results of the second part of the study (Kaplinsky et al. 2002; Metz and Fütterer 2002a).

The same critics obscure the validity of the primary finding of the paper by focusing on the flaws in the second, exploratory section and by failing to distinguish adequately between the two independent but related aspects of Quist and Chapela's work—the detection of transgenes in maize landraces and the effort to begin analyzing the behavior of the transgenes in this new genomic context. The critics did not effectively contest the first finding, which was later confirmed, as we noted earlier. Their responses focused heavily on the exploratory and tentative results of the second section of Quist and Chapela's report. Even while it is tentative, the effort in the second section is not quite as faulty as the critics depict it to be. Metz and Fütterer claim that "their i-PCR products all seem to be artifacts of the methodology used" (Metz and Fütterer 2002a, 600). Kaplinsky and coauthors likewise suggest that all eight sequences derived from i-PCR are artifacts of false priming—that is, mistakes produced by errors in research technique—yet one of the genetic sequences identified by Quist and Chapela (the sequence labeled AF434761) did retrieve a substantial amount (more than thirty base pairs) of nonprimer CaMV sequence and thus cannot be labeled an artifact. This contradicts the critics' assertions that Quist and Chapela's i-PCR work did not produce any meaningful results. Instead of dismissal and hyperbolic condemnation, a more reasonable response from critics would have been to suggest reasons why this valid i-PCR result defied certain expectations, along with what should be standard practice in such cases of uncertainty—calls for further investigation, such as that articulated by Quist and Chapela themselves in their initial report.

Even with its problems, the work discussed in the second part of Quist and Chapela's paper serves as a place holder for a critical research area: the determination of what happens with transgenes in uncontrolled ecological contexts. Such an important scientific question is unlikely to be answered adequately on the first attempt, so criticism of early work is expected. The reactions to Quist and Chapela's article, however, seem to have become suspended in an extended criticism phase, far beyond the criticisms published in *Nature*, with little interest expressed for more investigation into transgene behavior in the environment (an opportunity missed in *Nature*'s editorial note, which accompanied Quist and Chapela [2002] and simply distanced the journal from its original decision to publish the article). Intensive investigation of the behavior of transgenes is clearly needed: Even the molecular mechanism of transformation (the way in which transgenes become inserted into host genomes) remains uncertain (Granger 2002), and thus the stability of transgenic constructs in uncontrolled environmental samples cannot be assumed.

The implication of this uncertainty is that current and historic introductions of transgenic organisms into the environment, which Quist and Chapela's findings suggest cannot be contained, amount to a global gene transfer experiment with unknown and unpredictable ecological consequences. Reversing this trend requires more research on ecological genomic phenomena in situ (for instance, transposons—DNA sequences that can move around to different positions in an organism's genome, sometimes causing mutations), the behavior and effects of transgenes or trans-DNA, and the relative importance of these when transgenic methods are applied to agricultural, social, economic, and cultural systems and problems. Quist and Chapela have contributed in this regard. Uncertainty will always be present, but further research and observation may provide a more balanced view in both scientific and policy circles.

The scientific debate surrounding Quist and Chapela's findings takes place within webs of political and financial influence that compromise the appearance of objectivity (and perhaps the actual objectivity) of scientists such as the Quist and Chapela critics (Kaplinsky et al. 2002; Metz and Fütterer 2002a). The majority of the authors of the two critiques of Quist and Chapela published by *Nature* recently had all or part of their research funded by the Torrey-Mesa Research Institute (TMRI), a progeny of the ag-biotech firm Novartis, now operating under the name Syngenta. The affiliation of many of those authors with TMRI was a result of that company's $25 million "strategic alliance" with the College of Natural Resources at the University of California, Berkeley. (According to the agreement with the University of California, Novartis/Syngenta received first right to negotiate for research results from the Department of Plant and Microbial Biology and strong representation on the committee that disbursed the funds to laboratories in the department.) Wilhelm Gruissem, formerly of Berkeley and the architect of the strategic alliance, was the supervisor of another of the authors, Johannes Fütterer. In publishing their reactions to Quist and Chapela's research, none of these authors declared this funding from an ag-biotech firm as a competing financial interest.

Such a funding arrangement might be less noteworthy had Chapela not been the leading faculty critic, and Quist a leading student critic, of the Berkeley-TMRI strategic alliance and its implications for scientific freedom and balanced science. Both Quist and Chapela were then and still are affiliates of Berkeley, as were most of their primary published critics, and as are we. All this made Berkeley the nexus of interlinked controversies surrounding university-industry relations and transgenes in Mexican maize. Quist and Chapela's vocal opposition to the alliance jeopardized a substantial amount of industry research money for some of these same scientists who, out of the thousands of biotechnology researchers qualified to evaluate Quist and Chapela's research, went on to become their chief critics. These competing interests would be less striking, furthermore, had some of these critics not resorted to publicly accusing Quist and Chapela of incompetence and ideological bias; Metz called their paper a "testament to technical incompetence" and suggested that "an ideological conflict encouraged this lapse in scientific integrity" (quoted in Yoon 2002). Given the history of the Berkeley-TMRI controversy, one wonders why *Nature* did not look further afield for critics of Quist and Chapela, rather than choose directly from the center of that controversy for both critiques that it published.[3]

Compromised positions extend beyond these critics. Nature Publishing Group, the publisher of *Nature, Nature Biotechnology,* and a host of other scientific journals, actively integrates its interests with those of companies invested in agricultural and other biotechnology, such as Novartis, AstraZeneca, and other "sponsorship clients," soliciting them to "promote their corporate image by aligning their brand with the highly respected *Nature* brand."[4] These conflicts of interest challenge *Nature*'s ability to provide a neutral forum for scientific debates on ag-biotech. *Nature*'s editorial note was unorthodox and unnecessary—the normal scientific process of contestation could have been permitted to proceed, using Quist and Chapela's claims and data to repeat, verify, or refute their findings, without additional editorial comment.

Because of its potential effect on regulatory policy, the timing of *Nature*'s action further undermines the journal's quest to be perceived as uncompromised by commercial and financial interests. The editor's disavowal of Quist and Chapela's original report and publication of critical responses occurred immediately before the sixth meeting of the United Nations Environment Programme Convention on Biological Diversity and discussions of the legally binding Cartagena Protocol on Biosafety (April 7–19 and 22–26, 2002, respectively). The protocol "aims to ensure the safe transfer, handling and use of living modified organisms that result from modern biotechnology that may have adverse effects on biological diversity" (United Nations Environment Programme 2002). In light of these aims, the debate about the flow and possible fragmentation of transgenes was particularly important. *Nature's* potential influence on the biosafety meeting became a reality when the delegate from Australia invoked *Nature's* editorial note, weakening the power of Quist and Chapela's findings (see Ho 2002).

Commercial connections are not unusual for scientific journals and their editors and authors, as elucidated by Sheldon Krimsky and others quoted in *Nature's* news feature on conflicts of interest (Van Kolfschooten 2002). Unfortunately, peer reviews and requirements for disclosure of competing financial interests provide scant counterbalance to the webs of interests surrounding ag-biotech science. Direct influences are only part of that environment. Biotechnology, it must be noted, represents a prominent and seductive pathway for biological scientists into industry and toward financial and professional rewards. Current, outright competing financial interests, even when disclosed, do not sufficiently account for the progressive levels of interest that many scientists have in ag-biotech, when they work for ag-biotech firms, hope to do so in the future, are funded by those firms, or are ideologically committed to the proposition that industrial solutions are essential for solving agricultural problems. The extensive presence of industry in private and public research laboratories biases many laboratories against producing work that could be perceived as jeopardizing the future of ag-biotech. The pervasiveness of biotechnology advertisements in *Nature's* pages is one striking sign of the contemporary intensity of the marriage of science to corporations. Some people who perceive a lack of research that challenges the progress of ag-biotech in the pages of such prestigious scientific journals (such research is certainly not lacking in social science publications) look to industrial influences as a contributing factor. Scientists committed to the goal of impartiality cannot dismiss the effects of such influences on the signers of AgBioWorld's "Joint Statement in Support of Scientific Discourse in Mexican GM Maize Scandal" (AgBioWorld 2002b). To illustrate the point, all seventeen of the

Berkeley researchers who signed that statement were funded at the time by the ag-biotech firm TMRI, as described earlier. (See Department of Plant and Microbial Biology 2002 for the Berkeley-TMRI funding allocations.)

The pervasiveness of industry and financial biases certainly places the idea of completely eliminating them beyond the realm of possibility. Disclosure policies are created with the intention of mitigating the most obvious of those influences by allowing readers to make informed judgments about their significance. Yet closer examination of the financial interests and choices of *Nature* and its Quist and Chapela critics reveals a breakdown even in this limited protection. *Nature*'s policy defines *competing financial interests* as "those of a financial nature that, through their potential influence on behavior or content or from perception of such potential influences, could undermine the objectivity, integrity or perceived value of a publication" (Nature Publishing Group 2002b). The policy defines multiple ways that researchers may be linked financially to companies that stand to gain or lose from their work and publication, including support for research, equipment, supplies, and so on; recent, present, or anticipated employment; and personal financial interests such as stocks in a company. The authors of the second critique, Kaplinsky and colleagues, who actively declared that they had no competing financial interests (as opposed to making no declaration, which *Nature* leaves as an option), stand in clear violation of *Nature*'s guidelines on at least one count: coauthor Michael Freeling's research laboratory was funded by the ag-biotech firm TMRI, as noted earlier; beyond that, at least four of the six authors do research in laboratories funded by TMRI (or did at the time their critique was published), and some were graduate students whose education was partially funded by TMRI. In our judgment, critics Metz and Fütterer also violate the spirit of *Nature*'s policy because of the recent undeclared funding of Fütterer's research by the ag-biotech firm Novartis (see the acknowledgements in Rothnie et al. 2001). Metz also benefited recently from TMRI funding as a Berkeley graduate student and has been an outspoken proponent of the Berkeley-TMRI strategic alliance, as demonstrated by his testimony before the California legislature (Metz 2000; see also Metz 2001). Determining whether there were any other competing financial interests among these authors would require additional investigation.[5]

Also troubling is *Nature*'s failure to follow its own disclosure policy as it applies to publishing. *Nature*'s stated policy is to disclose its commercial or financial interests or specific arrangements with advertising clients or sponsors when such arrangements create "any risk of a perception of compromise in publishing and editorial decisions" (Nature Publishing Group 2002b). *Nature* publishes many ag-biotech advertisements, and historically its "sponsorship clients" include firms with ag-biotech interests, such as Aventis (maker of "StarLink" corn) and Novartis (although these companies' turbulent genealogies make tracking their affiliations challenging). These clients stand to lose from the findings in Quist and Chapela's paper because the results call into question the idea that ag-biotech firms can safely control transgenes in the environment—a challenge that raises the potential for increased regulation of transgenic organisms. *Nature*'s decision to publish a potent disavowal of Quist and Chapela's findings, without attendant disclosure of its financial relationships to ag-biotech firms, has created actual—not just potential—perceptions of compromise on the issue.

In such an environment, it is difficult to imagine fair and equal consideration being given to work that challenges the commercially vested interests of ag-biotech and the assumptions of traditional reductionist approaches to molecular biology. Quist and Chapela's paper obviously represents such a challenge. That fact—not the quality of their work— together with the politics of university-industry relations, is central to their paper's troubled reception. Ironically, the ag-biotech industry ultimately undermines its own credibility by not aggressively evaluating the health and environmental implications of its products. The public will remain skeptical until it does so. We call on scientists, *Nature*, and other scientific journals to reexamine their commitments to and interests in ag-biotech and to open up spaces in laboratories, journals, conferences, and classrooms for a more balanced and critical evaluation of the ecological and health effects of the flow of transgenes into the environment.

Acknowledgements

We are grateful to Eric Dubinsky and Lara M. Kueppers, whose expertise, insights, advice, and editorial feedback were vital to the drafting of this chapter. Thanks also to Shana Cohen for reviewing earlier drafts of this piece.

Notes

Portions of this chapter were published in the "Correspondence" section of *Nature* on June 27, 2002 (Worthy, Strohman, and Billings 2002).

The Berkeley Biotechnology Working Group includes Jason A. Delborne, Earth Duarte-Trattner, Nathan Gove, Daniel R. Latham, and Carol J. Manahan.

1. Quist and Chapela first reported contamination of Mexican maize landraces by transgenes in *Nature* (Quist and Chapela 2001). *Nature* subsequently published two severe critiques of that report (Kaplinsky et al. 2002; Metz and Fütterer 2002a) followed by a response from Quist and Chapela (2002) in which they provided additional data supporting their original finding of contamination. Together with these, the editors of *Nature* added an unusual "Editorial Note" that attempted to distance the journal from its decision to publish Quist and Chapela's original report (2001). See the editorial note accompanying Quist and Chapela (2002). Remarkably this action by the editors of *Nature* seems to be without precedent: "For the first time in *Nature*'s 133-year history, the journal had withdrawn support for an article without first calling for a retraction" (Platoni 2002). The original report, the critiques and the "Editorial Note" took center stage in the international controversy (or scandal) addressed in this chapter. In Worthy, Strohman, and Billings (2002), we disclosed the initial results of our investigation into the webs of political and financial influence challenging the supposed neutrality and objectivity of the critics of Quist and Chapela that *Nature* chose to publish and of *Nature* itself. Alongside that piece, *Nature* simultaneously (and with no prior disclosure to us) published replies to our claims from those critics (Kaplinsky 2002; Metz and Fütterer 2002b), in addition to a second unorthodox and perplexing editorial "comment" on the journal's original decision to publish the report, including a declaration of the importance of neutrality for the journal. Worthy (2002) provides a subsequent reply to the critics.

2. Some delays in confirming the presence of transgenes in the maize landraces stemmed from the proprietary nature of many transgenes. Researchers had difficulty accessing the exact specifications of the transgenes that they were investigating. Therefore they searched for other signs of genetic engineering, such as CaMV (a promoter sequence used in the vast majority of transgenic crops to enhance the expression of transgenes inserted beside it).

3. Chapela's ordeal has unfortunately extended far beyond the attacks and other difficulties mentioned above. Since September 2001, he has been immersed in an extremely protracted and difficult tenure review process that, as of this writing (August 2004), stands in a state of appeal of denied tenure, with possible legal action on the horizon. This is the case in spite of overwhelming support for Chapela's tenure from the Berkeley faculty and external reviewers. As reported in a breaking news story in *Nature*, a newly released review by Berkeley's Academic Senate concludes that "Jasper Rine, a geneticist at the university who sat on a key committee reviewing Chapela's tenure, had conflicts of interest. It says that Rine had financial dealings with biotech firms, oversaw the Syngenta

agreement, and had cited Chapela's *Nature* paper as an example of poor science in one of his classes. Both the dean of Chapela's college and his department chair requested that Rine be taken [removed] from the committee four times; but Rine did not excuse himself nor did the committee chair ask him to leave." The report further concludes that there was "unjustifiable" delay in Chapela's tenure review process (Dalton 2004, 598). These same conclusions are further supported by a newly released, commissioned external review of the Berkeley-TMRI agreement by researchers at Michigan State University, which devotes an entire section to the negative impact of the agreement on Chapela arising from his role as a leading critic of it. The external review states, "Regardless of whether Chapela's denial of tenure was justified, there is little doubt that the UCB-N [a.k.a. Berkeley-TMRI] agreement played a role in it." (Busch et al. 2004, 41, 42) Both the *Nature* news report and the MSU external review tie the injustices experienced by Chapela in his tenure review to his outspoken criticism of the Berkeley-TMRI agreement as well as to the controversy surrounding his *Nature* paper on Mexican maize contamination.

4. See Nature Publishing Group (2002a). Note that since the original publication of our Correspondence in *Nature* regarding these conflicts of interest (Worthy et al. 2002), Nature Publishing Group has removed the language about leveraging its "brand" and "corporate image" to attract biotechnology sponsorship clients.

5. The critique authors attempt to refute our claims of conflicts of interest, stating flatly that they had none. See Kaplinsky (2002) and Metz and Fütterer (2002b). Responses to them can be found in Worthy (2002).

References

AgBioWorld. 2002a. Mexican Maize Resource Library. http://www.agbioworld .org/biotech_info/articles/mexmaizeresource.html (accessed June 4, 2004).

———. 2002b. Joint statement in support of scientific discourse in Mexican GM maize scandal, February 24. http://www.agbioworld.org/jointstatement.html (accessed June 4, 2004) .

Busch, Lawrence, Richard Allison, Craig Harris, Alan Rudy, Bradley T. Shaw, Toby Ten Eyck, Dawn Coppin, Jason Konefal, Christopher Oliver, with James Fairweather. 2004. *External review of the collaborative research agreement between Novartis Agricultural Discovery Institute, Inc. and the regents of the University of California.* East Lansing, MI: Institute for Food and Agricultural Standards, Michigan State University. http://www.berkeley.edu/news/ media/releases/2004/07/external_novartis_review.pdf (accessed August 5, 2004).

Brown, Paul. 2002. Mexico's vital gene reservoir polluted by modified maize. *Guardian* (London), April 19. http://www.guardian.co.uk/gmdebate/Story/ 0,2763,686955,00.html (accessed June 4, 2004).

Dalton, Rex. 2004. Review of tenure refusal reveals conflicts of interest. *Nature*

430 (August 5): 598. http://www.nature.com/cgi-taf/DynaPage.taf?file=/
nature/journal/v430/n7000/full/430598a_fs.html (accessed August 5, 2004).

Department of Plant and Microbial Biology. 2002. Funded research projects
(2002–2003). http://plantbio.berkeley.edu/~pmbtmri/Projects.html (accessed
June 4, 2004).

Enciso L., Angélica. 2002. Confirma el INE la presencia de transgénicos en cul-
tivos de Oaxaca. *La Jornada* (Mexico City), August 12. http://www.jornada
.unam.mx/2002/ago02/020812/042n1soc.php?printver=1 (accessed June 4,
2004).

Granger, Claire. 2002. Transgenes by no easy means. *Information systems for
biotechnology news report*, February. http://www.isb.vt.edu/news/2002/
news02.Feb.html (accessed June 4, 2004).

Ho, Mae-Wan. 2002. "Worst ever" contamination of Mexican landraces. *ISIS Report*, April 29. http://www.i-sis.org.uk/contamination.php (accessed June 4,
2004).

Kaplinsky, Nick. 2002. Correspondence: Conflicts around a study of Mexican
crops. *Nature* 417 (June 27): 898.

Kaplinsky, Nick et al. 2002. Biodiversity (Communications arising): Maize transgene results in Mexico are artefacts. *Nature* 416 (April 11): 601–2.

Metz, Matthew. 2000. Testimony before the California Senate, Senate Natural
Resources Committee/Senate Select Committee on Higher Education. Impact of genetic engineering on California's environment: The role of research
at public universities. Sacramento, May 15.

———. 2001. . . . But Syngenta deal is a boon to Berkeley. *Nature* 410 (March
29): 513.

Metz, Matthew, and Johannes Fütterer. 2002a. Biodiversity (Communications
arising): Suspect evidence of transgenic contamination. *Nature* 416 (April
11): 600–1.

———. 2002b. Correspondence: Conflicts around a study of Mexican crops. *Nature* 417 (June 27): 898.

Monbiot, George. 2002. The fake persuaders: Corporations are inventing people
to rubbish their opponents on the Internet. *Guardian* (London), May 14.
http://www.guardian.co.uk/Columnists/Column/0,5673,715158,00.html (accessed June 4, 2004).

Nature Publishing Group. 2002a. Partnerships. http://npg.nature.com/npg/
servlet/Content?data=xml/10_sponsor.xml&style=xml/10_sponsor.xsl (accessed May 6, 2002; Nature Publishing Group has since changed the text of
this page).

———. 2002b. Statement of policy on competing financial interests. http://www
.nature.com/nature/submit/competing/index.html (accessed June 4, 2004).

Platoni, Kara. 2002. Kernels of truth. *East Bay Express*, May 29. http://www
.eastbayexpress.com/issues/2002-05-29/feature.html/1/index.html (accessed
June 4, 2004).

Quist, David, and Ignacio H. Chapela. 2001. Transgenic DNA introgressed into traditional maize landraces in Oaxaca, Mexico. *Nature* 414 (November 29): 541–43.

———. 2002. Biodiversity (Communications arising [reply]): Suspect evidence of transgenic contamination/maize transgene results in Mexico are artifacts. *Nature* 416 (April 11): 602.

Rissler, Jane, and Margaret Mellon. 1996. *The ecological risks of engineered crops.* Cambridge, Mass.: MIT Press.

Rothnie, Helen M., Gang Chen, Johannes Fütterer, and Thomas Hohn. 2001. Polyadenylation in rice Tungro bacilliform virus: Cis-acting signals and regulation. *Journal of Virology* 75 (May): 4184–94.

Shouse, Ben. 2002. Corn controversy heats up. *ScienceNow,* April 4. http://sciencenow.sciencemag.org/cgi/content/full/2002/404/2?ck=nck (accessed June 4, 2004).

United Nations Environment Programme. 2002. Governments to advance work on Cartagena protocol on biosafety. Press release. http://www.biodiv.org/doc/meetings/bs/iccp-03/other/iccp-03-pr-en.pdf (accessed June 4, 2004).

Van Kolfschooten, Frank. 2002. Conflicts of interest: Can you believe what you read? *Nature* 416 (March 28): 360–63.

Worthy, Kenneth. 2002. Responses to Metz, Fütterer and Kaplinsky's correspondence in *Nature,* 27 June 2002. http://nature.berkeley.edu/~kenw/maize/responses.htm (accessed June 4. 2004).

Worthy, Kenneth, Richard C. Strohman, and Paul R. Billings. 2002. Correspondence: conflicts around a study of Mexican crops. *Nature* 417 (June 27): 897. http://www.nature.com/cgi-taf/DynaPage.taf?file=/nature/journal/v417/n6892/full/417897b_fs.html (accessed June 3, 2004).

Yoon, Carol Kaesuk. 2002. Journal raises doubts on biotech study. *New York Times,* April 5.

9

Hard Red Spring Wheat at a Genetic Crossroad

Rural Prosperity or Corporate Hegemony?

R. Dennis Olson

No man qualifies as a statesman who is entirely ignorant on the problems of wheat.

Socrates

Overview

Shortly after Monsanto submitted its 2002 applications to deregulate its Roundup Ready hard red spring wheat in both the United States and Canada, the company publicly pledged that it would not commercially release the world's first strain of genetically engineered wheat until several conditions were met. First, Monsanto pledged to gain market acceptance for genetically engineered wheat by convincing major international wheat buyers to agree to purchase it. Second, regulatory agencies in the United States and Canada would have to approve Monsanto's genetically engineered wheat simultaneously, so as not to give one country a market advantage over the other. Third, Monsanto pledged to work with the wheat industry and the regulatory agencies to establish "appropriate" contamination thresholds.[1] And finally, Monsanto committed to work with the wheat industry and the regulatory agencies to resolve unanswered questions about the costs of segregation and to address outstanding agronomic stewardship concerns (Monsanto Canada 2003). By March of 2004 Monsanto had failed to meet virtually all these self-imposed conditions, but nevertheless considered moving ahead with seeking approval of its genetically engineered wheat only in the United States, in direct contradiction to its public commitments (Reuters 2004a). After strong resistance to this move from the U.S. wheat industry, Monsanto first announced, on May 10, 2004, that it would discontinue funding for all research on genetically engineered wheat (Monsanto 2004). Then, on June 18, 2004, Monsanto announced that it was withdrawing its pending applications for regulatory approval of the genetically engineered wheat from all agencies from around the world except for the U.S. Food and Drug Administration (Reuters 2004b). This stunning retreat by a major biotechnology company from the marketing of a major biotech crop, even if it proves only temporary, represents an historical bellwether in the ongoing controversy over the safety of biotech crops for humans, biodiversity and rural economies.

The commercial release of genetically engineered wheat would have had a profound effect on farmers in North America, who played a pivotal role in pressuring Monsanto to abandon commercialization of genetically engineered wheat. Recent economic projections indicate that the commercial release of genetically engineered wheat would have caused a devastating collapse in prices that farmers get for their wheat. This was the primary reason why leaders within the North American wheat industry stubbornly questioned and effectively resisted Monsanto's bid to commercialize its genetically engineered wheat variety. This loss of farm income would have occurred because most major wheat importers have vowed not to buy genetically engineered wheat—because contamination of native wheat by genetically engineered wheat is inevitable—and the resulting loss of these export markets would have relegated North American wheat farmers to the role of supplier of last resort. Unlike genetically engineered corn and soybeans—which are used mostly for animal feed—genetically engineered wheat would have gone directly into the human food system. This would have likely meant even stronger consumer resistance to foods contaminated with genetically engineered wheat. The lesson of Monsanto's historic retreat is that policy makers the world over must carefully weigh the socioeconomic and agronomic ramifications of this new food crop before commercially and irretrievably releasing genetically engineered crops into farm fields and food distribution systems.

"Aristocrat" of Wheat at the Center of Genetic Engineering Controversy

Hard red spring wheat "stands out as the aristocrat of wheat for baking bread." It has the highest protein content of all U.S. wheat varieties and therefore has greater gluten content. Many flour mills—both in the United States and abroad—desire this characteristic so they can blend hard red spring wheat with lower-protein wheat varieties to increase the gluten content in the flour that they mill. In the United States, hard red spring (HRS) wheat is grown primarily in four states: North Dakota, South Dakota, Montana, and Minnesota (North Dakota Wheat Commission 2002a). It also grows on the Great Plains provinces of Canada.

In the winter of 2001, the North Dakota House of Representatives, citing the potential loss of wheat exports resulting from consumer rejection of genetically engineered foods, passed a temporary ban on the commercial introduction of genetically engineered wheat—a shot across the bow of the multinational Monsanto, which had announced plans to introduce its genetically engineered Roundup Ready strain of HRS wheat by the planting season of 2003. The shot got Monsanto's attention.

The company and its allies launched an intensive lobbying effort to kill the genetically engineered wheat moratorium in the North Dakota Senate. With the help of the powerful chair of the state Senate Agriculture Committee, Monsanto succeeded in getting the moratorium bill watered down to a nonbinding interim study resolution. However, the interim study helped keep the issue in the limelight for the next two years by mandating research and public hearings. Significantly, a wheat farmer who ran against the agriculture committee chair won in 2002 by making the state senator's opposition to the ban on genetically engineered wheat a key campaign issue (Gillis 2003).

In 2001, farmers promoted similar bills for a moratorium and an interim study of genetically engineered wheat in Montana, another major HRS wheat state. Monsanto and its allies killed both bills there as well. The debate about Monsanto's proposed introduction of genetically engineered wheat continued to escalate, with new legislation introduced again in 2003 in both North Dakota and Montana. The new legislation would have (1) required state certification of any genetically engineered wheat variety before its commercial release (North Dakota State Legislative Branch 2003a; Montana State Legislative Branch 2003a); and (2) changed state liability laws to apply the tenets of strict liability to genetically engineered seed producers in the case of contamination from genetically engineered wheat crops, thereby providing more protection to farmers not growing genetically engineered wheat from unfair liability issues stemming from alleged patent violations (North Dakota State Legislative Branch 2003b; Montana State Legislative Branch 2003b). However, all these bills were defeated as well. Similar genetically engineered wheat certification bills were introduced, but not passed, in South Dakota (South Dakota Legislature 2003) and Kansas (Kansas State Legislature 2003/2004) in 2003.

North of the border, opposition to genetically engineered wheat also began to grow in Canada. The Canadian Wheat Board, with strong support from wheat-producer organizations, conducted surveys of its foreign buyers of wheat (Canadian Wheat Board 2001); commissioned agronomic studies of the potential ramifications of genetically engineered wheat (Van Acker, Brule-Babel, and Friesen 2003); established minimum conditions before genetically engineered wheat could be introduced (Canadian Wheat Board 2003a); and proposed regulatory reform that would allow consideration of the potential loss of export markets in making a final decision to deregulate genetically engineered wheat (Canadian Wheat Board 2003b). Finally, the wheat board publicly called on Monsanto to withdraw its application for genetically engineered wheat deregulation (Canadian Wheat Board 2003c).

These challenges likely contributed to one of the first significant setbacks for Monsanto's effort to gain approval for its Roundup Ready wheat. In the spring of 2003 the U.S. Department of Agriculture (USDA) rejected Monsanto's pending application to deregulate its genetically engineered wheat because of deficiencies. The rejection meant that Monsanto would have to address deficiencies and resubmit a revised application. This development in turn meant that the regulatory review clock would be reset, and the agency would have another six months to review the application again, which is about the average time that USDA has taken to act on applications to deregulate other genetically engineered crop varieties (Center for Food Safety 2003b).

Roundup Ready Wheat: Contamination Is Inevitable

Objections to genetically engineered wheat are based on the belief that it will inevitably contaminate native wheat—restricting the ability of farmers, grain elevators, and processors to meet the requirements of customers who demand wheat free of genetic engineering. Proponents of genetically engineered wheat argue that it and natural crops can coexist with a dual marketing system that would solve the problem of consumer rejection by segregating the two crops. Such a system would provide genetically engineered U.S. wheat to buyers who accept it, and conventional and organic wheat to others who don't want it. The effectiveness of this approach is highly controversial, and the potential cost of a dual marketing system is open to debate (Wisner 2003, 22–25). Two recent cases indicate that segregating genetically engineered and non-genetically engineered crops is not sufficient to avoid serious contamination.

Mexican Corn Contamination

In October 2003, nongovernmental organizations (NGOs) released testing results that had detected genetically engineered contamination in nine Mexican states, despite Mexico's ban on the importation or cultivation of genetically engineered corn. Analyses of these tests revealed the presence of transgenes in native varieties of corn, including StarLink contamination, and the presence of genes in some plants from as many as four different genetically engineered varieties, all patented by transnational biotechnology corporations (Action Group 2003; see also Snow, Gepts, and Worthy, et al., this volume).

"This is just a small sample," stated Ana de Ita of the Center for Studies on Rural Change in Mexico (CECCAM), "but it indicates the seriousness of the problem. If we're finding contamination in random samples from indigenous and farming communities far from urban centers and in communities that have traditionally used their own seed, then the problem is much more widespread. . . . The plants in several communities that contain two, three and even four different transgenes together indicates that the contamination has been around for years, and that contaminated maize on small farms has been cross-pollinating for generations" (Action Group 2003).

Silvia Ribeiro of the Action Group on Erosion, Technology and Concentration, an NGO that supports socially responsible developments of technologies useful to the poor and marginalized and addresses international governance issues and corporate power, warned that "recent U.S. production of corn genetically modified to produce substances ranging from plastics and adhesives, to spermicides and abortifacients poses an even greater risk of contamination. There have already been cases in Iowa and Nebraska of accidental escape of corn modified to produce non-edible substances. If we're already finding contamination in remote areas of Mexico, where cultivation of GM corn is prohibited by law, how can we guarantee that these other types won't spread as well?" (Action Group 2003).

Canadian Contamination of Canola Seed Stock and Related Concerns about Roundup Ready Wheat

The contamination of seed stocks caused by transgene flow from Roundup Ready genetically engineered canola has devastated the organic canola industry in Canada, prompting the filing of a class action lawsuit by organic growers against Monsanto and Aventis, a European biotech company that introduced its own variety of genetically engineered canola. The lawsuit argues that genetically engineered canola has contaminated seed stocks, farm fields, and the canola distribution systems to such an extent that it has virtually wiped out the organic canola market for Saskatchewan farmers. "You can't grow organic canola in Canada anymore, simply because the GM variety exists," said Jim Robbins, a Canadian canola farmer (Reuters 2003). The suit seeks damages against the two companies, arguing that they were negligent for failing to implement effective measures to prevent farm-to-farm contamination (Organic Consumers 2002).

Because many organic canola growers depend heavily on wheat in their crop rotation, their lawsuit seeks an injunction to stop the commercial release of genetically engineered wheat as well. "We have lost canola as a crop in our rotations because of genetic contamination, but we obviously cannot afford to lose wheat which is our largest crop and largest market," said Arnold Taylor, president of the Saskatchewan Organic Directorate. According to a CNEWS article, Canadian wheat exports are estimated at $2.9 billion annually, and losing any fraction of that amount to market rejection of genetically engineered wheat by importing countries would be devastating, both to individual farmers and to the rural communities that depend on the wheat economy (CNEWS 2002).

The Canadian Wheat Board commissioned a study that assessed the factors contributing to contamination of the Canadian canola crop by genetically engineered canola and speculated about whether similar contamination would occur with the introduction of genetically engineered wheat (Van Acker, Brule-Babel, and Friesen 2003). Because many of the factors are the same, (Van Acker, Brule-Babel, and Friesen 2003, 15–16), the study concluded that the potential level of contamination of the wheat crop caused by the unconfined release of Roundup Ready wheat would be similar to the substantial contamination that occurred with canola (Van Acker, Brule-Babel, and Friesen 2003, 1).

However, the Van Acker study points out that the place to address the question of creating a viable dual segregation system must be the farm, not the grain elevator. "The segregation issue is inextricably linked to Roundup Ready wheat management issues because both are about limiting transgene movement from Roundup Ready wheat to non-Roundup Ready wheat," write Van Acker, Brule-Babel, and Friesen (2003, 23). The report concludes that any efforts by industry and government to segregate genetically engineered wheat are likely to fail, just like they failed in regard to canola:

Any effort made to keep the Roundup Ready trait discrete within Roundup Ready canola have proven insufficient, even in the pedigreed seed production systems which can be considered an intensive segregation system. Given the similarities between wheat and canola with respect to a Roundup Ready transgene bridge it is likely that current commercial and seed production management systems in wheat would be insufficient to keep the Roundup Ready trait discrete within Roundup Ready wheat. *Management systems sufficient to achieve and maintain discrete segregation of the Roundup Ready trait in either wheat or canola have not yet been devised, modeled or tested.* (17; emphasis added)

Loss of Export Markets

The most compelling threat fueling wheat farmers' resistance to Monsanto's introduction of its genetically engineered wheat is the loss of export markets. Across the northern Great Plains, wheat farmers have begun to ask tough questions about Monsanto's Roundup Ready HRS wheat, which would be the first genetically engineered wheat variety ever released commercially. Of major concern are the substantial, unanticipated adverse economic effects from lost export markets. This happened when consumers rejected genetically engineered varieties of other major commodities, such as corn, soybeans, and canola. U.S. government officials have estimated that the European Union's regulations for genetically engineered crops could cost the United States as much as $4 billion in agricultural exports to Europe alone (Pew Initiative on Biotechnology 2002, 9).

Rejection of Genetically Engineered Wheat by Asian and European Buyers

Asia and Europe are major customers for both U.S. and Canadian HRS wheat. In the 2000–2001 marketing year, six of the top ten importers of HRS wheat were Asian countries, and two were European (North Dakota Wheat Commission 2002b). Substantial consumer resistance to genetically engineered foods in these countries ranges from labeling requirements to calls for outright bans on the importation of genetically engineered crops (International Forum 2003).

Private wheat-processing companies are taking even stronger stands against genetically engineered crops than their governments. For example, the Japanese Millers Association, which mills 90 percent of Japanese wheat, has stated that it will not purchase wheat that has any level of contamination by genetically engineered wheat (Northern Plains Resource Council 2002). Mandatory labeling of genetically engineered food means that many consumers are still likely to reject these products even with government approval of "acceptable" tolerance levels. A September 2002 survey found that 100 percent of Japanese, Chinese, and Korean wheat buyers responding to the poll said that they would not buy genetically engineered wheat under any circumstances, even if their governments gave regulatory approval for it (Western Organization 2003a).

Similarly strong resistance to genetically engineered crops exists in Europe. Peter Jones, an official with Rank Hovis, which controls more than 30 percent of the milling and baking in the United Kingdom, said: "I am going to ask you not to grow genetically modified wheat until we are able to sell in our market the bread made from the flour made from that [GE] wheat. I cannot tell you how to run your business—but if you do grow genetically modified . . . wheat, we will not be able to buy any of your wheat—neither GM nor the conventional" (Western Organization 2003a). Nicolaas Konijneenkijk, president of the Dutch company called AGRO Consulting and Trading, summed up the general attitude among

European wheat buyers when he said, "Wheat and bread are sacred in Europe and many other parts of the world. If farmers and government officials in the U.S. fail to recognize that, they can kiss their markets good-bye" (Western Organization 2003a).

Worsening Erosion of U.S. Wheat Export Market Share

In assessing the threat of potential export losses from genetically engineered wheat, it is important to consider the intensified competition for international markets that has undermined U.S. dominance in wheat exports since the early 1980s. The U.S. market share of world wheat exports has fallen from a peak of nearly 50 percent in the 1970s to a low approaching 20 percent in 2001 (Wisner 2003, 26). The USDA predicts that various factors will continue to erode the U.S. share of the wheat market for the foreseeable future (U.S. Department of Agriculture 2002, 6).

The introduction of genetically engineered wheat is likely to exacerbate this bleak outlook for U.S. wheat exports: If the United States attempts to supply world markets with both genetically modified and unmodified wheat from the same producing regions, foreign consumers will be faced with the following question: *"Should I buy non-GMO U.S. wheat at a premium price that includes the costs of segregation and certification, or can I get similar wheat from other suppliers without paying segregation costs?"* (Wisner 2003, 26). Major wheat-importing countries such as Japan that wish to secure alternative sources of unmodified wheat outside the United States would have incentive to make substantial investments in regions—including several former Soviet republics—that have the capacity to increase their wheat production to meet this new demand.

Economic Costs of Genetically Engineered Wheat to Farmers and the Rural Economy

In an October 2003 study, Dr. Robert Wisner—a leading grain market economist at Iowa State University—looked specifically at the potential ramifications that Monsanto's pending commercial introduction of genetically engineered HRS wheat would have on U.S. export markets and the wheat economy. Key findings of the study include:

- Thirty to 50 percent of the foreign market for U.S. hard red spring wheat, and an even greater percentage of U.S. durum wheat exports, could be lost if genetically engineered HRS wheat is introduced in the United States before 2010.

- Average U.S. prices for HRS wheat would be forced down to feed-wheat price levels, approximately one-third lower than the average of recent years.
- Durum and white wheat exports and prices also would likely face substantial risk; other classes of wheat would face slightly lower risk.
- Loss of wheat export markets would lead to loss of wheat acreage; loss of revenue to industries supplying support services to wheat producers; losses for other rural farm-related and nonfarm businesses, local and state government tax revenues, and institutions supported by tax revenues; and diminished economic health of rural communities and state governments in the spring wheat belt (Wisner 2003, executive summary).

"A large majority of foreign consumers and wheat buyers do not want genetically modified wheat," Wisner concluded. "Right or wrong, consumers are the driving force in countries where food labeling allows choice" (Western Organization 2003b).

"While there are many unknowns about genetically modified wheat, one thing's for certain," said Helen Waller, a Montana wheat farmer. "Commercial introduction into Montana, North Dakota, and other wheat-producing states could result in our wheat commanding only feed grain prices, consequently reducing our market price by a third. And that will put farmers like myself out of business" (Western Organization 2003b).

In June of 2003, Monsanto publicly pledged that it would not introduce genetically engineered wheat until it is accepted in major markets. Nevertheless for nearly a year the company refused demands by the Canadian Wheat Board and others to withdraw its applications for approval of its Roundup Ready wheat (Monsanto Canada 2003). According to analysts, Monsanto is in a precarious financial situation (Innovest Strategic Value Advisors 2003). Additionally, the company now faces lawsuits from farmers, and allegations from former executives, that it engaged in price fixing and violated antitrust laws (Barboza 2004). Given the potential revenues from Roundup Ready wheat, Monsanto's unwillingness to delay its application was not surprising. In the end, however, the threat of international market rejection, and tough grassroots resistance both from farmers and consumers, finally compelled Monsanto to agree reluctantly to withdraw virtually all its regulatory applications for genetically engineered wheat pending at the U.S. Food and Drug Administration (Monsanto 2004).

Potential Effects of Genetically Engineered Wheat on the U.S. Domestic Market

U.S. Consumers Skeptical Too

Although much of the debate has focused on loss of overseas markets, there also are substantial concerns about how Monsanto's genetically engineered wheat introduction would affect the U.S. domestic wheat market. Polling has consistently shown that a large majority of U.S. consumers want genetically engineered foods to be labeled and that a substantial portion of the population would avoid genetically engineered foods if they were labeled. For example, an ABC News Poll in June 2001 found that 93 percent of Americans say that they want genetically engineered foods labeled. "Such near-unanimity in public opinion is rare," commented the poll's authors. Additionally, this poll found that "[b]arely more than a third of the public believes that genetically modified foods are safe to eat. Instead, 52 percent believe such foods are unsafe, and an additional 13 percent are unsure about them. That's broad doubt on the very basic issue of food safety" (Langer 2001).

Such a high level of skepticism toward genetically engineered food causes U.S. wheat processors to question the wisdom of blindly forging ahead with the approval of Monsanto's genetically engineered wheat deregulation. "In every study [of U.S. consumers] . . . there's still 7 to 10 percent of people who say 'I will not buy a product if it contains a genetically modified ingredient,'" stated Ron Olson, vice president of General Mills. "When you come to a company like ours, which is a wheat-based organization, and we run the risk that we will lose 7 to 10 percent of our business if you change a product and it becomes an issue . . . I don't think that's a risk our corporation would take" (Western Organization 2003a). However, it seems that Monsanto is counting on making a profit on genetically engineered wheat once government regulatory agencies approve "appropriate" contamination levels. Once genetically engineered wheat is released commercially, it will be impossible to segregate it from nongenetically engineered wheat, according to the study by Van Acker, Brule-Babel, and Friesen (2003). Eventually, it will be next to impossible to guarantee that any wheat is GE-free. Over time, that level of contamination will increase, and natural wheat will eventually disappear. Then buyers and consumers will have no choice but to buy products containing contaminated wheat and concerns about Monsanto's product will be a moot point.

It is not a lack of concern among consumers that has stifled debate in the United States about the safety of genetically engineered crops; it is rather the U.S. government's failure to allow a debate. It was in this vacuum that wheat farmers and legislators from the northern Great Plains in early 2003 joined with U.S. consumer groups to demand that the USDA conduct a complete environmental impact statement (EIS) on the effects of deregulating Monsanto's Roundup Ready wheat. The petition from farmers, legislators, and consumers contended, among other things, that the government must identify the potential socioeconomic, agronomic, and environmental impacts that would result if the agency were to approve Monsanto's application for genetically engineered wheat deregulation and develop means of mitigating the adverse effects of genetically engineered wheat (Center for Food Safety 2003a). The USDA may never respond to the petition because Monsanto has withdrawn its application, but the issues it raised may have contributed to the agency's decision to reject Monsanto's application as deficient.

Foreign Importers Could Win Increased Share of U.S. Domestic Wheat Market

Many U.S. wheat processors export finished wheat products to Asia and Europe, and there is concern that they could lose domestic market shares to foreign imports of wheat if genetically engineered wheat were introduced in the United States. Dan McGuire of the American Corn Growers Association explained this threat at a town hall meeting in Montana in January 2003 (before Monsanto withdrew its applications): "If U.S. farmers were to grow GMO wheat, U.S. millers might import conventional wheat from Europe and elsewhere so as not to jeopardize not only their market with U.S. consumers but also their market for flour and wheat products that they export from the U.S. to buyers around the world that won't accept products made from GMO wheat" (McGuire 2003).

Resistance Grows: Survey Reveals Grain Elevator Opposition to Genetically Engineered Wheat

In the spring of 2003, the Institute for Agriculture and Trade Policy, which promotes family farms, rural communities, and ecosystems around the world through research, education, science and technology, and advocacy, surveyed grain elevator operators in North Dakota about their views on Monsanto's pending application. Ninety-eight percent of the operators responding to the survey said that they were either very concerned (82 percent) or somewhat concerned (16 percent) about the proposed introduction of genetically engineered wheat. Additionally, 78 percent of the respondents supported an expanded public review of genetically engineered wheat, one that would go well beyond what the USDA has required for approval of genetically engineered crops in the past (Institute 2003).

"The world wide consumer must have confidence with the credibility of the U.S. farmer and government dealers, which will have no control should Monsanto be in control of wheat releases," said one elevator operator. "Where is the demand for Roundup Ready wheat? Not one consumer group wants it!" (Institute 2003).

"Release [of genetically engineered wheat] before customer acceptance could be death to the U.S. spring wheat market," said another operator. "It's impossible to have a segregation system with zero tolerance" (Institute 2003).

In other survey results, the North Dakota elevator operators ranked loss of export markets as their greatest concern related to genetically engineered wheat deregulation. All said that their customers were concerned about genetically engineered wheat deregulation, with 77 percent characterizing their customers' concern about genetically engineered wheat as either very high (54 percent) or high (23 percent) (Institute 2003).

Confirming the skepticism among North Dakota elevator operators, the National Grain and Feed Dealers Association recently estimated that fewer than 5 percent of U.S. grain elevators have the capacity to operate a dual grain-marketing system (Wisner 2003, executive summary).

Whose "Right to Choose" for Whom?

The Government now has all of the necessary information to make a sensible and definitive statement about the future of GM crops within the UK. It is time to move forward in a sensible, responsible manner that will give back UK growers and consumers the right to choose, whilst allowing the economy and the environment the opportunity to benefit from this exciting technology.

Dr. Paul Rylott, chair,
British Agricultural Biotechnology Council (2003)

Proponents of genetic engineering often accuse critics of genetically engineered crops of infringing on farmers' and consumers' "right to choose." However, evidence continues to mount that genetically engineered crops limit, and in some cases may even eliminate, the right of farmers to choose to raise unmodified crops—either conventional or organic—and sabotage the consumers' right to choose *not* to eat genetically engineered food. The cases of Mexican corn contamination and Canadian organic canola contamination are clear examples of this.

Proponents of genetically engineered crops also argue that society should protect the "right" to plant these crops, because genetic engineering is a management tool that may increase farmers' profits. There have been some documented cost savings for certain genetically engineered crops—especially for initial users. However, over time, because of the increase in herbicide-resistant weeds and other factors, studies show that the farmer's costs for growing genetically engineered crops rise substantially, calling into question the alleged economic benefits of such crops for individual farmers (Benbrook 2003).

Additionally, the Canadian Van Acker report breaks down the farm costs for both adopters of the genetically engineered technology and nonadopters. Just for managing Roundup Ready "volunteers"—seeds that germinate a year or more after the initial planting—the additional estimated on-farm costs for adopters range from $6.95 to $15.37 (U.S.) per acre for low-disturbance direct seeding to $1.46 to $3.66 (U.S.) per acre for high-disturbance direct seeding or conventional tillage. For nonadopters, the costs range from $5.49 to $11.47 (U.S.) per acre for low-disturbance direct seeding,[2] with no increased costs for high disturbance (conventional plowing) (Van Acker, Brule-Babel, and Friesen. 2003, 19).[3] It is one thing for proponents to argue that whether to adopt genetically engineered technology is the choice of the individual farmer, who also therefore then decides to incur the extra management burdens reflected in this report. However, since when does our society sanction the "right" of individual farmers to impose such clear and irreversible additional operating costs on her or his neighbors who choose *not* to use it?

Finally, proponents of genetically engineered wheat often tout the environmental benefits of these crops. However, the Van Acker report concludes that the unconfined release of genetically engineered wheat would cause the loss of reduced tillage cropping systems, which would in turn result in increased soil erosion, increased herbicide loads on ecosystems,

and increased greenhouse gases. Low disturbance tillage cropping systems do not rely on conventional plowing for weed control. These systems
leave crop residues as ground cover and rely as heavily on more intensive
chemical applications—like Roundup—to control weeds. This reduces
soil erosion—and is especially effective in the arid Great Plains. The Van
Acker study found that the introduction of genetically engineered wheat
would result in an increase the application of Roundup herbicide to control an expected increase in volunteer Roundup wheat prior to seeding in
low disturbance cropping systems. The study also projected that farmers
would have to increase in the application of other chemicals to control the
emergence of other types of Roundup resistance weeds. These increased
chemical applications required by the introduction of genetically engineered wheat would cause increased costs to all farmers—both adopters
and non-adopters. Not only would these increased chemical loads
threaten the environment, but when added to the chemical loads already
being applied in crop rotations for Roundup Ready canola they would ultimately threaten the long-term sustainability of reduced tillage cropping.
(Van Acker, Brule-Babel and Friesen 2003, 1–2; 25–26).

    The current U.S. policy of deregulating genetically engineered
crops—including Roundup Ready and other crops—poses a clear and
present danger to the integrity of seed stocks and biodiversity worldwide.
It threatens the very existence of conventional and organic crops, and it
threatens to eliminate the right of farmers to choose organic or conventional crops over genetically engineered crops—not to mention the right
of consumers to choose *not* to eat genetically engineered food. Because of
unwavering market rejection of genetically engineered wheat among
most international buyers, and because of dogged grassroots opposition
from farmers, rural communities and consumers, Monsanto has temporarily corked the genetically engineered wheat genie in the bottle. The
evidence now available overwhelmingly argues against the deregulation
and unconfined release of genetically engineered wheat, and even Monsanto has acknowledged that fact—if only indirectly through the withdrawal of its regulatory applications for genetically engineered wheat
around the world. Wheat farmers and the rural economies that depend
upon wheat cannot afford the economic catastrophe that would inevitably
result if this genie is unleashed before the myriad problems already documented with other crops are resolved. Our valuable wheat must be protected from irreversible genetic contamination.

Notes

1. The Webster's New World Dictionary defines the word "contaminate" as meaning: "to make impure, corrupt, etc. by contact; pollute; taint." Current and proposed threshold standards for permitting trace levels of genetically engineered materials in both conventional and organic crops, and for creating viable systems for segregating genetically engineered and non-genetically engineered materials, do not distinguish between contamination caused through gene flow, such as occurs with pollen drift, and contamination caused by simple physical mixing of genetically engineered seeds, harvested crops or processed flour with conventional or organic seeds, crops or processed flour. Therefore, throughout this paper I use the term "contaminate," and "contamination," to mean, interchangeably, either contamination by gene flow, or physical mixing of genetically engineered and non-genetically engineered seeds, harvested crops or processed products.

2. "Direct seeding, like no-till, is a cropping system which aims to improve soil and soil moisture conservation. Direct seeding is more flexible than no-till; it allows some tillage to solve immediate weed problems and to deal with high moisture and heavy clay soil conditions." Direct Seeding System: Terms, Definitions and Explanation. Agriculture and Rural Development website of the Alberta Government: http://www1.agric.gov.ab.ca/$department/deptdocs.nsf/all/agdex3483?opendocument (accessed on August 16, 2004).

3. The amounts cited here have been converted to U.S. dollars from Canadian dollars, the latter of which were used in the Van Acker study. I used the Money Converter program on the website titled, "Babel Monery Conversion Around the World," at: http://oanda.com/converter/classic?user=Babel, which I set for June 1, 2003 (accessed August 16, 2004). This date coincides with the date of publication of the Van Acker study of June 2003. The actual figures in the Van Acker study in Canadian dollars are: $9.50 to $21 per acre for high-disturbance direct seeding or conventional tillage, to $2–$5 per acre for high-disturbance or conventional tillage for adopters; and, the costs range from $7.50–$16 per acre for low-disturbance direct seeding for non-adopters.

References

Action Group on Erosion, Technology and Concentration. 2003. Nine Mexican states found to be GM contaminated. Press release from indigenous and farming representatives and from civil society organizations, October 11. http://www.etcgroup.org/article.asp?newsid=410 (accessed December 23, 2003).

Agricultural Biotechnology Council. 2003. GM crops can and should co-exist in the UK: It's time to give UK consumers and farmers the right to choose. Press release, November 25. http://www.bioportfolio.com/news/abc_2.htm (accessed December 24, 2003).

Barboza, David. 2004. Questions linger on price of seeds: Two companies said to have agreed to an increase. *New York Times* January 6, p. A1.

Benbrook, Charles. 2003. Impacts of genetically engineered crops on pesticide use in the United States: The first eight years. *Biotech Infonet.* Technical paper number 6, November. http://www.biotech-info.net/technicalpaper6 .html (accessed January 7, 2004).

Canadian Wheat Board. 2001. Grain matters. *CWB Publications,* September–October. http://www.cwb.ca/en/publications/farmers/sept-oct-@2001/09-10–01-3.jsp (accessed January 6, 2004).

———. 2003a. DRAFT 5 Discussion document: Conditions for the introduction of genetically modified wheat. Grain Industry Working Group on Genetically Modified Wheat, February 5. http://www.cwb.ca/en/topics/biotechnology/ gmowheat.jsp (accessed January 6, 2004).

———. 2003b. Closing the gap on GE wheat. Testimony of Canadian Wheat Board chair Ken Ritter, and farmer-elected CWB director Bill Nicholson before the House of Commons Standing Committee on Agriculture and Agri-Food, Ottawa, March 31. http://www.cwb.ca/en/topics/biotechnology/ closing_gap.jsp (accessed November 14, 2003).

———. 2003c. CWB asks Monsanto to put the brakes on Roundup Ready Wheat. CWB news releases, May 27. http://www.cwb.ca/en/news/releases/2003/ 052703.jsp (accessed on January 6, 2004).

Center for Food Safety. 2003a. Petition seeking an environmental impact statement concerning the deregulation of genetically engineered wheat varieties & petition seeking the listing of genetically engineered wheat varieties as noxious weed. March 11. http://www.centerforfoodsafety.org/li.html (accessed January 5, 2004).

———. 2003b. Background on determinations for "deregulated status." Factsheet. http://www.agobservatory.org/library/uploadedfiles/Background_ on_Determination_for_Deregulated_St.pdf (accessed March 30, 2004).

CNEWS. 2002. Saskatchewan organic farmers file lawsuit against Monsanto and Aventis. *Canadian Press,* January 10. www.canoe.ca/CNEWSDangerous-Foods0201/10_lawsuit-cp.html (accessed December 23, 2003).

Gillis, Justin. 2003. The heartland wrestles with biotechnology. *Washington Post,* April 22, p. A1.

Innovest Strategic Value Advisors. 2003. Monsanto & genetic engineering: Risks for investors. Report, April.

Institute for Agriculture and Trade Policy. 2003. New survey indicates strong grain elevator concern over GE wheat. Press release. Minneapolis, Minnesota, April 8.

International Forum on Globalization and the Center for Food Safety. 2003. Genetically modified [GM] crops and foods: Worldwide regulation, prohibition and production. Joint project map, July. http://www.centerforfoodsafety.org/ facts&issues/ElectronicVersion.pdf (accessed January 7, 2004).

Kansas State Legislature. 2003/2004. SB 236: An act relating to genetically modified organism crops. http://www.kslegislature.org/bills/2004/236.pdf (accessed January 7, 2004).

Langer, Gary. 2001. Behind the label: Many skeptical of bio-engineered food. ABCNEWS.com, June 19. http://abcnews.go.com/sections/scitech/DailyNews/poll010619.html (accessed November 14, 2003).

McGuire, Dan. 2003. GMO wheat . . . A market development-in-reverse program. Presentation to GE Wheat Forum, Montana Winter Fair, Lewistown, January 23.

Monsanto. 2004. Monsanto to realign research portfolio, development of Roundup Ready wheat deferred. Press release, Monsanto Website. http://www.Monsanto.com/Monsanto/layout/media/04/05-10-04.asp (accessed August 2, 2004).

Monsanto Canada. 2003. Monsanto restates pledge to meet all Roundup Ready wheat commitments. Press release, as edited and disseminated by AGNET (an on-line agricultural news service), Canada, June 25. http://www.gene.ch/genet/2003/Jun/msg00103.html (accessed January 5, 2004).

Montana State Legislative Branch. 2003a. HB 409: The Montana wheat protection and promotion act. http://leg.state.mt.us/css/default.asp (accessed January 6, 2004).

———. 2003b. HB 522: Revise liability for genetically engineered wheat. http://leg.state.mt.us/css/default.asp (accessed January 6, 2004).

North Dakota State Legislative Branch. 2003a. HB 2106: A bill to create a transgenic wheat board. http://www.state.nd.us/lr// (accessed January 6, 2004).

———. 2003b. SB 2304: A bill relating to damages for cross-pollination with transgenic wheat. http://www.state.nd.us/lr// (accessed January 6, 2004).

North Dakota Wheat Commission. 2002a. Wheat information: Hard red spring wheat, about hard red spring wheat. http://www.ndwheat.com/wi/hrs (accessed November 13, 2003).

———. 2002b. Top 10 markets for U.S. HRS & durum—2000–2001. Wheat information: Market information: Charts and stats: U.S. HRS & durum wheat. http://www.ndwheat.com/wi/markstat/top10_export_bushel.asp (accessed November 13, 2003).

Northern Plains Resource Council. 2002. Why plant what you can't sell? Protecting Montana's valuable wheat from genetic contamination. Factsheet. Billings, Montana, summer.

Organic Consumers Association. 2002. Canada organic farmers lawsuit against Monsanto moves forward: Organic farmers say report backs fears. *Leader-Post* (Regina, Saskatchewan), June 28. www.organicconsumers.org/gefood/organicfarmers602.cfm (accessed December 23, 2003).

Pew Initiative on Food and Biotechnology. 2002. U.S. v. E.U.: Examination of the trade issues surrounding genetically modified food. Brief, June (updated August 2003). http://pewagbiotech.org/resources/issuebriefs/europe.pdf (accessed June 4, 2004).

Reuters. 2003. Canadian growers warn UK farmers of GMO crop risks. November

3. http://www.planetark.com/dailynewsstory.cfm/newsid/22745/story.htm (accessed December 22, 2003).

―――. 2004a. Monsanto raises idea of U.S.-only wheat release, March 16. http://biz.yahoo.com/rc/040316/food_biotech_1.html (accessed March 22, 2004).

―――. 2004b. Monsanto withdraws GMO wheat from all but U.S. FDA, Friday June 18, 2004. Reuters.com (accessed Friday, June 18, 2004).

―――. 2004c. Monsanto withdraws biotech wheat petition, USDA. Reuters .com, March 11, 2004.

South Dakota Legislature. 2003. SB 214: An act to provide for the regulation of certain genetically modified wheat. http://legis.state.sd.us/sessions/2003/bills/SB214p.htm (accessed January 7, 2004).

Taylor, Michael R., and Jody S. Tick. 2003. Post-market oversights of biotech foods—Is the system prepared? A report commissioned by the Pew Initiative on Food and Biotechnology and prepared by Resources for the Future, Washington D.C., April.

U.S. Department of Agriculture Economic Research Service. 2002. U.S. wheat output & exports to decline in 2002/03. Commodity Spotlight; Agricultural Outlook, August. http://www.ers.usda.gov/publications/agoutlook/aug2002/ao293c.pdf (accessed January 4, 2004).

Van Acker, R. C., A. L. Brule-Babel, and L. F. Friesen. 2003. An environmental safety assessment of Roundup Ready Wheat: Risks for direct seeding systems in Western Canada. A report prepared for the Canadian Wheat Board for submission to the Plant Biosafety Office of the Canadian Food Inspection Agency, June.

Western Organization of Resource Councils. 2003a. Food safety: Buyers around the world reject genetically modified wheat. Food Safety section of website, May 23. http://www.safefoodfight.org/foodfight/ff_quotes.html (accessed January 7, 2004).

―――. 2003b. New study: The commercial introduction of genetically modified wheat would severely depress U.S. wheat industry. Press release, October 30.

Wisner, Robert. 2003. Market risks of genetically modified wheat: The potential short-term impacts of GMO spring wheat introduction on U.S. wheat export markets. Report prepared for the Western Organization of Resource Councils, Billings, Montana, October 30. http://www.agobservatory.org/library/uploadedfiles/Market_Risks_of_Genetically_Modified_Wheat.pdf (accessed June 4, 2004).

10

Agricultural Biotechnology and the Environmental Challenge

Peter H. Raven

From 1950, when the world was inhabited by about 2.5 billion people, to 2000, the population has more than doubled, to six billion. How can we sustain human life on this planet as the population continues to grow? Today, although poverty is declining steadily in Asia and Latin America, approximately 1.2 billion people are living on less than $1 per day; in sub-Saharan Africa, almost half the people have an income at or below that level. About eight hundred million people in developing countries are chronically undernourished, a reduction of approximately forty million since 1990 but still a very large number; worldwide, the World Health Organization estimates that about half the population is malnourished for at least one critical dietary element. From 1970 to 1999, average food consumption per person increased in all regions, from 2,100 to 2,700 calories in developing countries, and from 3,000 to 3,400 calories in industrialized ones. The world population may become stable at a level of approximately nine billion people during the course of this century, but even that conservative estimate, combined with expectations for more affluence and consumption everywhere, poses enormous challenges for the world agricultural system.

In an effort to supply these needs, about 11 percent of the world's land surface is used to produce crops, a collective area about the size of South America, and only limited potential remains for expanding the area of land under cultivation. Most of the additional gains will be made in South

America and in sub-Saharan Africa, and they will be made only with the full application of all the tools available to agriculture. At the same time, about 20 percent of the land that was arable in 1950 has been lost subsequently, to salinization, desertification, urban sprawl, erosion, and other factors, so that farmers are feeding 6.3 billion people today on about four-fifths of the land from which they were feeding 2.5 billion people in 1950. That is possible through a combination of selection, breeding, improved irrigation systems, soil conservation, and the judicious application of fertilizers. Modern agriculture scarcely resembles the agriculture of the 1940s, and yet it is not able, partly for political and social reasons, to feed all people well.

Beyond the land consigned to crop production, an additional 20 percent of the world's land surface is used for raising animals, very critical in a world that is increasingly shifting to animal proteins. Most land used for agriculture and grazing, especially in the tropics and subtropics, is being degraded by these activities and is therefore becoming less sustainable and productive in the face of increasing worldwide demand for high-quality food.

In the world as a whole, human beings are estimated to be using, wasting, or diverting nearly half the total products of photosynthesis, which is essentially the sole source of nutrition not only for humans but for all the other organisms on Earth. In addition, people are consuming more than half the total renewable supplies of freshwater in the world. Agriculture accounts for about 90 percent of the total water consumed for human purposes, and we are unlikely to choose to maintain this relationship indefinitely as our population grows.

In a world in which eight hundred million people receive so little food that their brains cannot develop normally and their bodies are literally wasting away; three billion people are malnourished; and in which 1.2 billion people live on less than $1 per day, the human population is supported by means of a gigantic and continuing overdraft on the world's capital stocks of water, fossil energy, topsoil, forests, fisheries, and overall productivity. We use the world, its soils, waters, and atmosphere as a gigantic dumping ground for pollutants, including the pollutants that render much surface water unusable, the carbon dioxide that is contributing directly to global warming, and the atmospheric pollution that kills millions of people around the world annually. We are not living sustainably, and we can clearly find our way to a sustainable future only by achieving a sustainable population, finding a sustainable level of consumption globally, accepting social justice as the norm for global development, and finding the improved technologies and practices that will help us make sustainable development possible.

One of the most serious pressures on the world's long-term sustainability is the loss of biodiversity. At present rates of destruction of the world's tropical moist forest, the UN has estimated that perhaps only 5 percent will remain by the middle of the twenty-first century. Such an enormous loss of habitat, according to the relationships demonstrated in the field of island biogeography, would in itself result in the loss of two-thirds of the species originally present in the areas as their ranges and population sizes become increasingly restricted. The loss of so many species clearly will have a negative effect on humans' prospects. We derive all our food, most of our medicines, and a major proportion of our building materials, clothing, chemical feedstocks, and other useful products from the living world. In addition, the communities and ecosystems that it comprises protect our watersheds, stabilize our soils, determine our climates, and provide the insects that pollinate our crops, among many other ecosystem services. Furthermore, we are early in the age of molecular discovery, when living organisms hold much of the promise for the development of currently unknown sustainable systems in the future. And, finally, by any moral or ethical standard, we simply do not have the right to destroy the diverse organisms that inhabit the planet, yet we are doing it savagely, relentlessly, and at a rapidly increasing rate, every day.

Among all human activities, agriculture, grazing, and forestry are the most destructive of biodiversity, accounting for the exploitative use of more than half the world's land surface. If we wish to stem the widespread extinction resulting from these practices, we must learn to make agriculture, grazing, and forestry as productive as possible on the lands that are being used; otherwise, a relatively unproductive, unfocused agriculture will lead to the destruction of many more species and more widespread destruction than would be the case if our existing systems were sustainable and as productive as possible. As Gordon Conway, president of the Rockefeller Foundation in New York, has pointed out, the single most promising way to avoid habitat destruction is to increase farm yields.

In the years since World War II, the application of relatively large amounts of synthetic pesticides has been essential to raising yields, but chemical agriculture has had highly negative environmental consequences.

We need ways to increase yields that are ecologically sound. Integrated pest management, involving the introduction of parasitic insects that would control pest species, is ecologically sound and widely applied. The practices followed in organic agriculture also contribute remarkably to the maintenance of soil fertility and the reduction of pesticide use and include many features that are of importance in the attainment of agricultural sustainability worldwide. However, organic farming is not necessarily sustainable and generally leads to reduced yields, according to long-term studies by the Swiss Research Institution for Sustainable Agriculture and comparable bodies. Organic agriculture is essentially what is practiced in sub-Saharan Africa today, and half the people are starving; so it is clear that organic agriculture by itself is not sufficiently productive to meet the world's food needs. Certainly, using less fertilizer and fewer pesticides in agricultural systems is in itself highly desirable, but productivity must be enhanced if even the huge areas now under cultivation are to meet human needs.

Where, then, do we go from here? A wide variety of new approaches will combine well to produce the more productive, sustainable agriculture of the future. What this new agriculture will look like will vary widely from region to region, and its attainment will require a high degree of imagination and a willingness to test many directions. To meet the real challenges of the intensive agriculture practiced widely in the modern world—and improve productivity and sustainability throughout—all available methods, certainly including biotechnology, must be applied where they will be useful.

The use of genetically modified organisms can improve the productivity and sustainability of agricultural systems. The use of technologies for genetic modification is important in achieving reductions in pesticide applications. Even by 2000, the use of genetically modified soybean, oilseed rape (canola), cotton, and maize had reduced pesticide use by 22.3 million kilograms of formulated product, and the reductions have gone far beyond that level subsequently. Worldwide, at least 500,000 people fall ill from pesticide poisoning each year, and five thousand die as a result. In the United States alone, approximately 110,000 cases of pesticide poisoning are reported each year, together with an estimated 10,000 cases of pesticide-induced cancer. Approximately 35 percent of the foods in supermarkets in the United States have detectable pesticide residues, which everyone would like to avoid. In the agricultural fields of the United

States, pesticides kill an estimated seventy million birds each year, as well as billions of both harmful and beneficial insects. Against this background, it is clear that the huge reductions already achieved in pesticide use constitute a major positive contribution to the environmental soundness of the agricultural systems in which these crops are being grown. These reductions in pesticide use provide a major benefit to the health of consumers wherever they have been attained.

Routine applications of pesticides in Europe are much higher than in the United States. It has been estimated that if half the maize, oilseed rape (canola), sugar beet, and cotton raised in Europe were genetically modified to resist their pests, European farmers would be able to reduce their use of formulated pesticide product by about 14.5 million kilograms, or 4.5 million kilograms of active ingredient. Using genetically modified crops would allow European farmers to reduce spraying by 7.5 million hectares, saving approximately 20.5 million liters of diesel and preventing the emission of 73,000 tons of carbon dioxide into the atmosphere. In the light of these figures, it is obvious that agriculture in Europe and throughout the world is being managed neither sustainably nor productively. In order to meet human needs adequately and safely, agricultural practices need to be improved everywhere.

Given the role that genetically modified crops can play in reducing pesticide use and raising agricultural productivity, why are these methods viewed with such skepticism? Other recent problems with food safety have contributed to the widespread public perception that genetically modified foods may somehow be unsafe in principle. In fact, no scientific theory exists that would explain why this should be so. Although people have consumed such foods in large quantities for many years, not a single case of sickness has been attributed to them. Many medicines, virtually all cheeses, much beer, and, in many regions, a large proportion of the other foods consumed have been produced as a result of the use of genetic modification. As a result, billions of people have been consuming huge amounts of genetically modified foods and medicines for many years: not a single case of sickness has resulted, and no scientific reason has been found that the process itself should be expected to produce deleterious effects.

If genetic modification itself were dangerous for some reason, people presumably would fear their doses of insulin, interferon, or other drugs produced through genetic engineering, regardless of how helpful they might be. But there is no such reason. In consideration of the facts, many learned bodies, including the Royal Society and the academies of science of many other countries, including the United States, China, India, Brazil, and Mexico, as well as the Third World Academy of Sciences and the Pontifical Academy of Sciences, have, considering the evidence, pointed out consistently over the years that there is no scientific basis for considering such foods unsafe for human consumption. Concerning food safety, it is time to stop dealing with phantoms and address reality for the benefit of human beings generally.

In the area of plant breeding, it is important to emphasize that when modern methods are used to study the production of individual characteristics in relation to the genes that carry these traits, the knowledge obtained can be used to improve crops all over the world, regardless of the commercial benefit obtained. It is general knowledge, of great common value. In contrast, traditional plant breeding improves only the specific crop involved, and the practice yields no general principles or specific facts that can be used for less profitable crops that may be essential to the livelihoods of hundreds of millions of people in developing countries. At the M. S. Swaminathan Research Institute in Chennai, India, for example, scientists for many years have been transferring genes for salt resistance from mangroves to rice in order to produce new strains of rice that can resist the brackish water infiltrating the coasts of India and still remain productive. These genes can in principle be used to improve the salt resistance of any crop, anywhere, and will be made available for that purpose.

In the face of widespread starvation in the developing world, activists who seek to persuade the governments of African countries, where hundreds of thousands of people are starving, to forgo food aid on the basis of politically or economically motivated disinformation seems to me to constitute a serious crime against humanity. I maintain that those responsible for the misinformation bear a serious responsibility for the lives of the people who are dying, and I urge the world as a whole to return to rationality in dealing with this humanitarian crisis. I see a serious moral dilemma that deserves greater consideration when citizens of industrialized countries accept medicines produced through genetic modification techniques while denying starving Africans foods that were produced in a similar way. As Per Pinstrup-Anderson, a senior research fellow of the International Food Policy Research Institute, points out, to a mother in a famine-struck region in Africa, she and her children suffer from a disease called hunger and the medicine is food. So the strong stand among U.S. and European activists and European governments against genetically modified crops, which have the potential to make more food available, may seem ill advised to hungry people in developing countries who need food—they don't need unsupported arguments about why it might not be safe. Serious discussions of the appearance of large-scale agriculture, the corporatization of food systems, or the globalization of trade are clearly necessary, but it is not genetically modified crops that are driving these trends, although activists sometimes use genetically modified crops as the poster child for these developments.

In 2002, the Congress of Racial Equality (CORE), one of the most venerable and respected civil rights groups in the United States, confronted Greenpeace at a public event and accused it of "eco-manslaughter" through its support of international policies limiting development and the expansion of technology to the developing world's poor. "But well-fed eco-fanatics shriek 'Frankenfoods' and 'genetic pollution,'" CORE said in a statement announcing its intention to confront Greenpeace. "They threaten sanctions on nations that dare to grow genetically modified crops, to feed their people or replace crops that have been wiped out by insects and blights. They plan to spend $175 million battling biotech foods over the next five years. Not a penny of this money will go to the starving poor." Although the amount may have been exaggerated, the principle seems right to me.

Apart from food safety, the fears concerning the cultivation of genetically modified crops are primarily environmental. Clearly, transgenes, like all the other genes that these crops possess, move regularly from crops to any wild or weedy relatives that may be growing in their vicinity. This process has been changing the characteristics of crops and their wild and weedy relatives since the beginnings of agriculture and is in fact a major feature of plant evolution in general. For some crops, such as maize and cotton in Europe, there are no wild or weedy relatives, and consequently there is no danger that any genes will spread. For others, such as oilseed rape (canola) and sugar beets in Europe, the consequences of any genes' moving into weedy populations, or genes' moving from the weeds into the crops, should be taken into account in deciding where to plant such crops. Whether a particular transgene persists in a natural population or not, and for how long, depends on its selective value in that population; the persistence of such a gene is by no means automatic. How would the role of the genetically modified weeds or wild plants differ from that of their unaltered relatives in agricultural systems or in nature, and would that constitute a problem? Are some of them likely to become weeds? Again, nothing intrinsic about the characteristics of the genetic modification process itself poses a threat. In the hands of those who wish to cripple the application of scientific techniques to the solution of human problems around the world, "gene transfer" has become a threat in itself and so emphasized as such that people have not paused to ask, "What is that threat?"

In conserving biodiversity throughout the world, we must improve the productivity and sustainability of all human activities, especially agriculture. Nothing has driven more species to extinction or caused more instability to the world's ecological systems than the development of agricultural practices sufficient to feed 6.3 billion people. The less focused and productive this agriculture is, the more destructive its effects will be. Measures of sustainability must be introduced widely to complement those of productivity, and forms of agriculture appropriate for individual regions must be designed and implemented. Drought and stress resistance, fewer additives of all kinds, improved productivity, and improved characteristics of the resulting foods must all be stressed in developing a wide variety of sound agricultural systems. Rational approaches to this field should lead gradually to the acceptance of genetic modification and other technologies and to their widespread use to help solve the many problems of agriculture. All parts of the process of acceptance by the public need to be transparent and verifiable, with questions addressed as they arise; only by a rigorous process of disclosure and investigation will a majority of people ever be comfortable with any new kind of technology.

New public sector efforts are needed to benefit poor farmers in developing countries, where in general neither the most important crops nor the conditions of cultivation have been the subject of much international effort. Whatever approaches might be taken to develop these agricultural systems, their precise modification by modern genetic techniques seems a rational way to move toward the desired outcomes. To assert that genetic modification techniques are a threat to biodiversity is to state the exact opposite of the truth. They and other methods and techniques must be used, and used aggressively, to help build sustainable and productive, low-input agricultural systems in many different agricultural zones around the world. Properly applied, they will provide major assistance for the preservation of biodiversity, and to the productivity and sustainability of ecosystems everywhere. They are, and will remain, an essential ingredient in building global sustainability.

In a fundamental sense, however, the only way to build a sustainable world is to change both that world and our way of thinking about it. A new industrial revolution and a new agriculture are clearly required to attain this goal. Population, overconsumption, and the use of appropriate technologies must all be brought into the equation to achieve it. Social justice must be extended to people everywhere. Their right to security, which underlies their ability to contribute to the formation of a world that will support both their children and ours, must be nurtured and expanded. In the words of Kai Lee, professor of environmental science at Brown University, we must continue to engage in a "search for a life good enough to warrant our comforts."

Note

This chapter is based on a lecture presented at the Natural History Museum, London, May 22, 2003, and sponsored by Sense about Science, a London-based charitable trust that encourages an evidence-based approach to scientific and technological developments.

PART 3

Hormone Replacement Therapy and Menopause: Science, Culture, and History

11

Postmenopausal Hormones

An Overview

Sylvia Wassertheil-Smoller

It is not often that a single study can overturn decades of thinking and have a profound effect on medical practice. However, that was what happened in the case of hormone replacement therapy. Usually, medical knowledge accumulates incrementally, but in this instance one large clinical trial, the Women's Health Initiative Clinical Trial of Estrogen Plus Progestin, upset all hopes and beliefs that hormone therapy for postmenopausal women was the key to health, beauty, and well-being. "I feel like I am a part of history," said Margaret, a participant in the study, after a television interviewer asked what she thought of having been among those taking estrogen plus progestin in the study. And indeed she is a part of medical history, as are the more than sixteen thousand women who also took part, as well as all the research investigators who conducted the Women's Health Initiative.

The Women's Health Initiative, known as WHI, was sponsored by the National Institutes of Health and consists of several interrelated studies whose overall objective is to prevent cancer, heart disease, and osteoporosis in postmenopausal women and to identify risk factors for these and other diseases of older women (Women's Health Initiative Study Group 1998). It is taking place in forty clinical centers throughout the United States, with 161,809 women participating nationally and being followed for ten to twelve years. One component of WHI consists of two trials of hormone therapy: a trial of estrogen alone for women who have had

181

a hysterectomy (known as the E-alone trial), and a trial of estrogen plus progestin for women who have not had a hysterectomy (known as the E+P trial). Standard care since the mid-1980s was to give women with a uterus progestin in addition to estrogen to protect the uterus from uterine (endometrial) cancer. The combination most widely used in the United States when the WHI began in 1993–94 was Prempro, manufactured by Wyeth-Ayerst. The trial of estrogen alone for women with a hysterectomy used Premarin. In both trials the comparison group took a placebo pill made to look exactly like the real thing. All study drugs were donated to the study by Wyeth-Ayerst.

Let us consider first what a clinical trial is and why we should do one. A randomized clinical trial is a prospective experiment to compare one or more interventions against a control group in order to determine the effectiveness of the interventions. Participants are randomly assigned to take the active drug, in this case, hormones, or the placebo. Randomization is done so that both groups are as similar as possible in all characteristics except for the pills that they are taking. Thus differences in outcomes between the two groups can be attributed to the treatment rather than to some baseline characteristic that may influence the outcome. Because many variables are unknown but may have a bearing on the results, randomization is insurance against unknown and unintentional bias. Randomization is intended to result in balanced groups, both with regard to variables that we know may influence outcome and variables that we may not even think of. In small clinical trials randomization may not always produce balanced groups, but randomization almost always works in large studies, and additional control for known unbalanced factors can be achieved in the statistical analyses. It should be noted that randomization in a clinical trial does not mean haphazard assignment; specifically, it means that each participant's chance of being assigned to the active pill group is the same as that of being assigned to the placebo group. Randomization is also the basis of the various statistical tests that we use to determine whether the results we get may be due to chance.

Whenever possible, clinical trials are double blind, meaning that neither the investigator nor the participant knows whether she is getting the active drug or a placebo pill. This is to avoid any possible systematic bias. The Women's Health Initiative Clinical Trial of Estrogen Plus Progestin (E+P) was a randomized, double-blind clinical trial of 16,608 women, with 8,506 assigned to take 0.625 mg conjugated equine estrogen plus 2.5 mg of medroxyprogesterone acetate daily, and 8,102 to take the placebo.

Why did we need such a trial in the first place? Hormones have been used in postmenopausal women for decades. The Federal Drug Administration (FDA) approved conjugated equine estrogen in 1942 for menopausal symptoms such as hot flashes and night sweats. In the 1970s, however, studies showed that unopposed estrogen was associated with uterine cancer, and the FDA issued a warning about this. In the 1980s studies found that progestin offset the risks of estrogen for uterine cancer, and doctors subsequently prescribed estrogen together with progestin for postmenopausal women who had a uterus. Estrogen alone was being prescribed for women who had had a hysterectomy and did not need the uterine protection. In the 1980s and 1990s, studies indicated that estrogens were probably effective against the bone loss that occurs with menopause, and in 1994 the FDA approved the use of estrogen plus progestin for the prevention of osteoporosis. In the meantime, however, studies were accumulating that suggested that estrogen offered protection against heart disease, and doctors increasingly prescribed hormones for this reason.

When the WHI began, six million women were taking estrogen and progestin, but no one knew the overall long-term risks of hormone therapy. Previous research had indicated that hormones reduced heart disease and osteoporosis, but these results came from observational studies. In such studies the investigator starts with a cohort of disease-free people who have a particular factor of interest, in this case, hormone taking, and people without that factor (the nonhormone-taking group), and goes forward into some future time to determine the frequency of development of disease (for example, heart disease) in the two groups. One potential source of bias in observational studies is that women who take hormones (or who have been prescribed hormones by their physicians) are different from women who don't. Compared to nonusers, hormone users are generally less obese, less likely to smoke, or to have diets high in fat and salt, more likely to be highly educated, physically active, go to doctors regularly (and thus to have their cholesterol and blood pressure monitored), and have mammograms and other screening tests. Such differences in the characteristics of the two groups, rather than the hormone taking, could explain why the hormone users appeared to have a lower risk of heart disease. Controlling for these factors in a statistical analysis may not be sufficient and may miss other, unknown factors that spell a difference between users and nonusers of hormones.

Often, clinical trials confirm evidence from observational studies, and numerous observational studies before WHI started indicated that postmenopausal women who take estrogen have a 30 to 50 percent lower rate of heart disease than those who don't take estrogen. Perhaps the best-known such study is the Nurses' Health Study (Stampfer et al. 1991), which first reported in 1991, before the WHI started, that post-menopausal women who had no history of cardiovascular disease and were currently taking estrogen alone had a 44 percent reduction in major coronary disease compared to those not taking estrogen, after adjustment for age and other risk factors. There was no association of estrogen alone and stroke. Fewer studies of estrogen combined with progestin were available at that time, but many of those also suggested benefits. However, a later report from the Nurses' Health Study, with a twenty year follow-up, of 70,533 was published in 2000, when the WHI was already under way. This report found that the risk for major coronary events among users of a combination of estrogen plus progestin was 36 percent lower than among women who had never used any hormones (Grodstein et al. 2000). But, there were no data from clinical trials to support this, and the FDA, citing insufficient evidence, turned down a request by Wyeth-Ayerst in 1990 to include a claim in the Prempro package insert that it prevented heart disease. Nevertheless, the belief in the observational study evidence was so strong that doctors were widely prescribing hormones not only for the indications approved by the FDA, osteoporosis and symptom relief, but specifically to prevent heart disease. As I noted earlier, only a ran-domized clinical trial could answer the question of long-term risks and benefits of hormone therapy, so the scientific community and women everywhere were looking forward to the results of WHI.

The WHI clinical trial of E+P was halted prematurely on July 9, 2002, on the recommendation of the Data and Safety Monitoring Board (DSMB). The DSMB is a group of independent experts from several dif-ferent disciplines who are not connected with the study and so have no conflicts of interest. The members see all results and all data—including which group is taking what—and are responsible for monitoring the safety of the volunteers taking part in the trial and for monitoring the quality of the trial. The DSMB examines the data at preset intervals and determines whether the trial should continue. In this case, the board determined that the E+P trial should be stopped because it showed a clear increase in breast cancer among women taking E+P compared with the placebo, and the overall harm outweighed the benefit. These results, and the reasons for stopping the trial, were rapidly published in the *Journal of the Ameri-*

can Medical Association (Writing Group for the WHI 2002), with subsequent articles coming out as additional analyses became available.

The release of the findings that estrogen plus progestin for postmenopausal women does more harm than good was startling, unexpected, and disturbing news to millions of women. The press and other media extensively reported the decision to halt this trial, which raised an outcry from women who thought that they were on a life-saving drug, as well as from the gynecologists who had widely prescribed it. The news also affected the stock market, sending shares of Wyeth-Ayerst, the manufacturer of Prempro, down by 24 percent on the first day. Women who had been on this drug were confused, because for years the evidence pointed to a protective effect with regard to heart disease, and they felt misled, angry, and perhaps even cynical about medical research and its mixed messages. Such a reaction is largely the result of some very common misunderstandings about the nature of scientific evidence, as well as our natural inclination to want definitive answers that support what we would like to believe. The good news for women now is that they have valid information that they can use to make their own decisions about what risks they are willing to tolerate for what benefits.

What, then, are the risks and benefits of E+P that were found in the WHI? The finding that led directly to the stopping of the trial was a 24 percent increase in invasive breast cancer. The higher risk of breast cancer with E+P was not surprising. Observational studies had indicated an elevated risk and in fact that possibility was spelled out in the consent form that the participants signed before they were randomized. However, the surprise was that the breast cancers found in the E+P group were larger and at a more advanced stage than those found in the placebo group (Chlebowski et al. 2003). Previously, the medical establishment had thought that cancers occurring in women on hormones would be smaller and at an earlier stage than those occurring in women not on hormones, but the WHI data showed the opposite. In addition, at every year of the study, the E+P group accounted for more abnormal mammograms than the placebo group. There was also a suggestion of increased ovarian cancer in the E+P group, although this finding did not reach statistical significance (Anderson et al. 2003). The biggest surprise, however, was the finding with regard to heart disease. WHI found a 24 percent increase in coronary heart disease, which included heart attacks, both fatal and nonfatal, as well as silent heart attacks found on electrocardiograms (Manson et al. 2003). The increased risk was most apparent at the one-year mark, when the increase in coronary heart disease was 81 percent. This, after all, was the primary aim of the study—to determine whether long-term hormone use prevents heart disease. It was totally unexpected to find that this hormone preparation actually increased the risk of heart disease.

There were compelling reasons for the earlier belief in prevention, one that was bolstered by a study called PEPI (Writing Group for the PEPI Trial 1995)—which stands for the Postmenopausal Estrogen/Progestin Interventions and was published in 1995. This was a randomized clinical trial that compared several hormone regimens to placebo: estrogen alone, estrogen in combination with progestin in the same preparation used in WHI, and estrogen plus a micronized progestin, which is a different kind of progestin. All the estrogen groups showed a decrease in low-density lipoprotein (LDL), which is directly related to heart disease risk, and an increase in high-density lipoprotein (HDL), which is the part of total cholesterol that is protective. Thus one would expect that a regimen that has a favorable effect on lipids would have a favorable effect on the endpoint of heart disease. That this was not confirmed in WHI brings up the very important point that a treatment that works on surrogate endpoints, such as lipids, may not work on the real endpoint of interest (in this case heart disease) that is related to the surrogate. The only way to determine whether hormones prevent heart disease is to determine the effect of hormones on heart disease itself, not on the intermediate factors of lipids, and that is what WHI was intended to do.

While WHI was in progress, the Heart and Estrogen/progestin Replacement Study (HERS) published results in 2001 from its randomized, blind, placebo-controlled trial of E+P in 2,763 women who already had established coronary heart disease, to see whether the hormone combination could prevent another heart attack. This is known as a secondary prevention trial, in contrast to primary prevention of heart disease in women who do not yet have coronary heart disease. The average age was sixty-seven, whereas in WHI it was sixty-three. The HERS investigators found more coronary heart disease events in the hormone group in the first year and fewer in the fourth and fifth years, with an overall null effect (Hulley et al. 1998). But there was still hope that the primary prevention tested in WHI, where 95 percent of the women had no prior history of heart attack, revascularization, or stroke (Stefanick et al. 2003), might work for women who were generally healthy and free of heart disease and that any early harm might turn into a long-term benefit, a hope that was dashed by further follow-up in HERS and that failed to materialize in WHI.

The WHI E+P trial also showed the expected increase in blood clots: venous thrombosis and pulmonary embolism (a twofold increase in risk with E+P; Writing Group for WHI 2002), as well as an increased risk of ischemic stroke (Wassertheil-Smoller et al. 2003). Ischemic stroke, caused by clotting in the brain, was 44 percent more likely in women on E+P than in those taking the placebo. An even more surprising finding came from the WHI Memory Study (Shumaker et al. 1998), which is an add-on study to WHI in which thirty-nine of the forty clinical centers participated. This study, known as WHIMS, was funded by Wyeth-Ayerst. The objective was to determine whether hormones protect against cognitive decline and dementia. Earlier evidence from animal and some human studies appeared to indicate the estrogen was protective of the brain, which has estrogen receptors. Not much research had been done on progestins, so little was known about the combination of estrogen and progestin. The WHIMS enrolled women who were already in the clinical trials of hormones and who were sixty-five or older. Astonishingly, the women on E+P had more than twice the rate of dementia (both Alzheimer's dementia and vascular dementia) as those in the placebo group, and there was no indication of prevention of cognitive decline with aging in general (Shumaker et al. 2003; Rapp et al. 2003).

So far we have been talking about relative risk, which means the risk of an event in one group relative to the risk of the event in another group. In follow-up studies such as the WHI, it is often expressed as a hazard ratio. A hazard ratio of 1.44 for E+P (as was observed for ischemic stroke, for example), means that women taking E+P are 44 percent more likely to have a stroke relative to those taking the placebo. For dementia, the hazard ratio was 2.05, meaning that women sixty-five and older who were taking E+P were 105 percent more likely to develop dementia than those taking the placebo (another way to say this is that they were 2.05 times as likely to develop dementia). This is a very useful way to describe the data when we are interested in the causes of disease, but for an individual woman who wants to assess her risks, it is more helpful to talk in terms of absolute risk and of excess absolute risk related to the intervention.

Let us take the case of breast cancer. There were 199 cases of invasive breast cancer in the E+P group and 150 cases in the placebo group. The

Table 11.1   Risks of E+P
Absolute Risks per 10,000 Women per Year

Event	E+P	Placebo	Excess risk E+P minus Placebo
Coronary Heart Disease (myocardial infarction, silent MI, CHD death)°	39	33	6
Stroke—all (Ischemic stroke only)°	31 (26)	24 (18)	7 (8)
Pulmonary embolism	16	8	8
Invasive breast cancer°	41	33	8
Dementia (ages 65+)°°	45	22	23

°Final data after 5.6 years of follow-up
°°Average follow-up 4.1 years

relative risk (or hazard ratio) was 1.24, but the absolute risk shows that for every 10,000 women aged 50 to 79 who were taking E+P for a year, there were 41 cases of invasive breast cancer, whereas for every 10,000 women taking placebo, there were 33 cases of invasive breast cancer, resulting in 8 *excess* cases per year per 10,000 women on E+P per year. So for any individual woman who is taking E+P, we cannot tell whether her breast cancer resulted from the hormones because, after all, many women not on hormones also got breast cancer. But on a population basis we can say that E+P will result in an excess of 8 cases of breast cancer per 10,000 women taking the preparation. For dementia, for 10,000 women aged 65 and older, there were 44 cases in the E+P group and 21 per 10,000 in the placebo group, or 23 excess cases annually for every 10,000 women 65 and older who were taking estrogen plus progestin. Table 11.1 shows the absolute risks and benefits for other endpoints in WHI per 10,000 women per year.

We've talked about the risks, but what were the benefits of E+P? There was a 24 percent reduction in all fractures and a 33 percent reduction in hip fractures (Cauley et al. 2003), as had been hoped and expected. E+P also reduced colorectal cancer by 37 percent (Writing Group for the WHI 2002). Table 11.2 shows the absolute rates for the conditions that showed benefit and the number of cases per 10,000 women per year that were *avoided* by the use of E+P.

Table 11.2   Benefits of E+P
Absolute Risks Per 10,000 Women Per Year

Event	E+P	Placebo	Excess Benefit E+P minus Placebo
Colorectal Cancer	10	16	–6
Hip Fractures° (total fractures)	11 (152)	16 (199)	–5 (–47)

°Final data

When we balance out the risks and benefits however, it is apparent that the adverse effects outweigh the beneficial effects. It is now clear, and the labeling required by the FDA so states, that estrogen plus progestin should not be used for the prevention of cardiovascular disease.

After these results came out, the medical community and women without a uterus were hoping that estrogen alone might have a better effect than estrogen combined with progestin, and everyone eagerly awaited the results of the estrogen-alone trial in the WHI. Another shock came on March 1, 2004, when participants received letters from NIH and from the principal investigators of the clinical sites, asking them to stop taking their study pills and to enter a follow-up phase. The estrogen-alone trial was stopped a year earlier than planned because it showed an increase in stroke risk and because an additional year's continuation would not change the conclusions reached. After an average of nearly seven years, estrogen alone (at a daily dose of 0.625 mg conjugated equine estrogen, Premarin) did not affect risk of heart disease or of colorectal cancer, decreased risk of hip and other fractures, and did not increase breast cancer; in fact there were fewer breast cancers in the estrogen alone group than in the placebo group, though this was not statistically significant. The risk of stroke, however, was increased with estrogen alone and was similar to that found in the study of estrogen plus progestin when that trial was stopped in 2002. Estrogen alone showed a 39 percent increase in strokes over placebo. In terms of absolute risk, there was an excess of 12 additional strokes per 10,000 women taking conjugated equine estrogen per year, compared to placebo (Women's Health initiative Steering Committee 2004). In addition, for women 65 years old or older, estrogen alone increased the risk of being diagnosed with either probable dementia or mild cognitive impairment by 38 percent: 93 of the 1,464 women taking estrogen alone compared to 69 of the 1,483 women taking placebo tablets, were diagnosed with either probable dementia or mild cognitive impairment (Shumaker et al. 2004). Estrogen alone also had a deleterious effect on overall global cognitive functioning. The adverse effect on cognition was greater among women who had lower cognitive function to begin with before treatment started (Espeland et al. 2004). These effects of estrogen and estrogen plus progestin on the brain are different from what would be expected based on animal studies and other basic science studies of the neurological effects of estrogen. Some people think that may be related to the timing of the administration of hormones. They suggest that if hormones are given to women just at menopause, rather than at older ages, they may be protective. There currently is no clinical trials evidence of whether this is so or not.

The results of the estrogen alone and the estrogen plus progestin trials are different in some ways and similar in others. They are similar in findings about hormones and the brain—both estrogen alone and estrogen plus progestin have adverse effects on the brain with regard to stroke and cognition. They are similar with regard to the protective effects on fractures. They differ with regard to heart disease: E+P increases risk, E alone does not affect heart disease. And they differ in their effects on both breast and colorectal cancer: E+P increases breast cancer while E alone does not; E+P protects against colorectal cancer, while E alone had no effect. It would be tempting to think that the differences are due to the progestin part of the hormone treatment, however, it is important to remember that the women enrolled in E+P were different from the women in the E alone trial. The women in the E+P had a uterus while those in E alone did not. Their baseline characteristics were different. So it is not clear yet why there were differences in effects. However, the conclusion that hormones should not be used for primary prevention of chronic disease applies to women with and without a uterus.

Despite the adverse effects, many women, and some physicians, have the impression that the quality of life is substantially better for women taking hormones. And some women say that are willing to take the risks in exchange for that presumably improved quality of life. What is the reality in this area? First, by definition, quality of life is a very subjective thing, but there are ways to describe it somewhat objectively. WHI investigators looked at thirteen measures of the following components of health-related quality of life: perceived health and physical functioning, bodily pain, mental health and depression, energy level and fatigue, social functioning and limitations on role activities, cognitive functioning, sleep, and sexual satisfaction. The researchers compared the E+P group and the placebo group for changes in these dimensions from baseline to one year on the study drug. They found statistically significant differences between the groups that favored E+P in three dimensions: body pain, physical functioning, and sleep, but these are considered small effects. After women were on the study drug for three years, the researchers could observe no differences among them. On the other hand, among women reporting moderate or severe vasomotor symptoms, for example, hot flashes or night sweats, all improved after one year, but a greater proportion on E+P showed improvement than among those on the placebo. Of those on the hormones, 71 to 77 percent reported improvement, compared to 52 percent among the placebo group. Among the youngest women in WHI, those aged fifty to fifty-four and thus closest to menopause, and who had moderate or severe vasomotor symptoms at baseline, only one of the thirteen measures showed statistically significant improvement with E+P and that was sleep. So, on average, E+P did not contribute in any important way to quality of life, although E+P undoubtedly helps some women with severe vasomotor symptoms. On a population basis, however, how large is that problem? At the time of screening for WHI, before they were enrolled in a particular part of the study, women were asked about a wide range of symptoms, including how bothersome hot flashes or night sweats were during the past four weeks. About 17 percent reported they had mild hot flashes, but only 7 percent reported moderate or severe hot flashes, with a similar percentage for night sweats, and only about 1.5 percent reported that these symptoms were severe.

Surely some placebo effect is operating in making women believe that the hormones improve their quality of life. For example, when the trial was stopped, before we told each participant which treatment she was on, we asked her what she thought she had been taking for the past several years, placebo or the active hormones. In our own clinic in New York City, of the first 445 women given the information, a little more than half, 62 percent, guessed correctly which group they had been in. Women who had been taking the hormones were more likely to guess that they had been on the hormones—72 percent of the 225 on hormones guessed correctly; however, of the 220 women who had been on the placebo, only 52 percent guessed that they had been taking the placebo, about equivalent to the toss of a coin. We asked the nurses, who did not know the treatment assignment of the participants, the same question. Amazingly, they agreed with the participant's guess less than a third of the time and with the true treatment assignment only 36 percent of the time, even though they had the participant's chart in front of them. These data show that the WHI staff was truly blinded.

This study was funded by NIH, and Wyeth-Ayerst provided drugs for WHI and fully funded the dementia study. The drug maker had, of course, hoped for positive results, and its officials were shocked and surprised that the results were so negative. It should be clear to everyone that this rigorously conducted study, the rapid dissemination of its results, and the multiple publications, which are still coming out as more analyses are being completed, were in no way influenced by the partial support of the pharmaceutical company.

The response of some gynecologists has often reflected wishful thinking. One claim is that the WHI was not a primary prevention trial because it included women who already had heart disease. In fact, less than 5 percent of the women had any previous history of stroke or heart attack, and the results did not change when the analyses were limited only to those women with no previous history.

Another issue that arises concerns the 40 percent of women who stopped taking their study pills sometime during the course of the trial. However, the percentage that stopped was very similar in both groups—42 percent in the E+P group and 38 percent in the placebo group. Thus the adherence rate does not invalidate the comparison between groups. Additionally, such an adherence rate is substantially higher than what occurs in real life—it has been estimated that as many as 80 percent stop taking hormones within the first four years of prescription (Pilon, Castilloux, and LeLorier 2001). Mainly, this objection is a misunderstanding about the intention-to-treat analysis. In clinical trials, the primary analysis should always follow the intention-to-treat principle, which shows that data from clinical trials in general should be analyzed by comparing the groups as they were originally randomized and not by comparing with the placebo control group only those in the drug group who actually did take the drug. The people assigned to the active drug group should be included with that group for analysis, even if they never took the drug. This may sound strange: How can one assess the efficacy of a drug if the patient isn't taking it? But the very reason that people may not comply with the drug regimen may have to do with adverse effects of the drug; therefore, if we select out only those who do comply, we have a different group from the one randomized, and we may have a biased picture of the drug's effects.

Those who adhere to a drug regimen in general may have some quality that affects outcome. A famous example of misleading conclusions that could arise from not doing an intention-to-treat analysis comes from the Coronary Drug Project (1980). This randomized, double-blind study compared the drug clofibrate to a placebo for reducing cholesterol. The outcome variable was five-year mortality and was very similar in both groups: 18 percent in the drug group and 19 percent in the placebo group. It turned out, however, that only about two-thirds of the patients who were supposed to take clofibrate actually were compliant and took their medication. Comparing the people in the drug group who actually took clofibrate to the placebo group showed that they had a significantly lower mortality—15 percent versus 19 percent in the placebo group. However, further analysis showed that among those assigned to the placebo group, only two-thirds took the placebo pills. The two-thirds of the placebo group who were compliant had a mortality of 15 percent, just like the ones who complied with taking clofibrate. The noncompliant people in both the drug and placebo groups had a higher mortality than the compliers (25 percent for clofibrate noncompliers and 28 percent for placebo noncompliers).

The inclusion of noncompliers in the analysis dilutes the effects, so every effort should be made to minimize noncompliance. In some trials, a judged capacity for compliance is an enrollment criterion, and an evaluation of whether this patient is likely to adhere to the regimen is made of every patient in determining his or her eligibility. Those not likely to comply are excluded before randomization. In the WHI, there was a run-in period before randomization to determine whether the participant was likely to take her pills, and nonadherence during the run-in period was a criterion for exclusion.

In WHI, the primary analysis was intention to treat. However, a secondary analysis adjusted for adherence. In these analyses, the event history of the participant was censored six months after she either stopped taking the study pills or was taking less than 80 percent of the study pills. In the placebo group, the event history was censored six months after the participant started taking hormones (some participants in the placebo group stopped taking study pills but were prescribed hormones by their physicians and started taking them on their own).[1] Thus this secondary analysis basically compared the two groups "as treated" or "per protocol," rather than as assigned to a particular treatment. In the intention-to-treat analysis, the hazard ratio for stroke, for example, was 1.31, while in the "as treated" analyses it was 1.50. In other words the findings from the intention-to-treat analysis were confirmed and strengthened in the adherence-adjusted analyses for stroke, coronary heart disease, breast cancer, and venous thrombotic embolism.

So far, in all the analyses done, the investigators have not been able to identify any subgroup of women for whom the risks and benefits differed from those for the group as a whole. However, WHI investigators continue to analyze the data to shed light on the biological mechanisms involved. In fact, a series of biomarker studies is being conducted on baseline blood drawn from women who went on to develop the various diseases under investigation. This blood was frozen and stored and is now being analyzed for biomarkers such as inflammatory and thrombotic factors, as well as genetic factors, and compared to blood drawn from women who did not develop the diseases under investigation. It is hoped that such analyses might better identify women who might be more likely to either benefit or have adverse effects from E+P.

Is there any place for hormone therapy for postmenopausal women? The FDA has approved use of this combination therapy for women who have a uterus and require treatment for moderate or severe vasomotor symptoms (hot flashes or night sweats), vulvar or vaginal atrophy, or prevention of osteoporosis (although this indication states that nonestrogen treatments to prevent osteoporosis should be considered first).

While the WHI results can make no statements with regard to the effects of other preparations and other routes of administration, such as patches, in view of WHI findings, the burden of proof of safety is on those other preparations. The FDA-approved statement in the boxed warning on the package insert for all estrogen and estrogen/progestin products for postmenopausal women states:

Other doses of conjugated estrogens and medroxyprogesterone acetate, and other combinations of estrogens and progestins were not studied in the WHI, and in the absence of comparable data, these risks should be assumed to be similar. Because of these risks, estrogens with or without progestin, should be prescribed at the lowest effective doses for the shortest duration consistent with treatment goals and risks for the individual woman.

The FDA has recently approved a combination therapy that uses lower doses of estrogen (0.45 mg compared with the 0.625 in Prempro) and progestin (1.5 mg rather than the 2.5 in Prempro).

What, then, is the net result of the WHI, and what does it mean for the postmenopausal woman? For the first time in the decades since hormones have been prescribed to millions of women, we have solid evidence (meaning rigorous clinical trial evidence) of its risks and benefits. For most women, the excess risks of heart disease, stroke, dementia, and breast cancer will outweigh the benefits of prevention of colorectal cancer and osteoporotic fractures and will be sufficient reason not to take these preparations. Some women who have severe menopausal symptoms may choose to take hormones. But these will be informed choices. The important thing to remember is that we are not talking about treating a disease with hormones; in the arena of treatment, one might be inclined to incur somewhat more risk to gain the benefit of cure or improvement in the condition. But menopause is not a disease. In weighing whether to take hormones or not, we are talking about preventing disease. And so the Hippocratic dictum is most relevant:

"First do no harm."

Notes

Any opinions expressed in this article are those of the author and do not necessarily represent those of other WHI investigators.

1. The Cox proportional hazards analysis is a statistical technique that uses the data from a person until the time that person is lost to follow-up or dies, when we say the data for that person is censored at the time the person was last known to be alive. Censoring for non-adherence means that six months after the person stopped being adherent (i.e., taking 80 percent or more of their study pills) it was treated as if the person were lost to follow-up—that is, the data from that person was no longer used in the analysis.

References

Anderson, Garnet L. et al. 2003. Effects of estrogen plus progestin on gynecologic cancers and associated diagnostic procedures: The Women's Health Initiative Randomized Trial. *Journal of the American Medical Association* 290:1739–48.

Cauley, J. A. et al. 2003. Effects of estrogen plus progestin on risk of fracture and bone mineral density: The Women's Health Initiative Randomized Trial. *Journal of the American Medical Association* 290:1729–38.

Chlebowski, Rowan T. et al. 2003. Influence of estrogen plus progestin on breast cancer and mammography in healthy postmenopausal women: The Women's Health Initiative Randomized Trial. *Journal of the American Medical Association* 289:3243–53.

Coronary Drug Project Group. 1980. Influence of adherence to treatment and response of cholesterol on mortality in the Coronary Drug Project. *New England Journal of Medicine* 303:1038–41.

Espeland, M. A. et al. 2003. Effect of estrogen plus progestin on global cognitive function in postmenopausal women: The Women's Health Initiative Memory Study: A randomized trial. *Journal of the American Medical Association* 289:2663–72.

Espeland, M. A. et al. 2004. Conjugated equine estrogens and global cognitive function in postmenopausal women: Women's Health Initiative Memory Study. *Journal of the American Medical Association* 291:2959–68.

Grodstein, F. et al. 2000. A prospective, observational study of postmenopausal hormone therapy and primary prevention of cardiovascular disease. *Annals of Internal Medicine* 133:933–41.

Hulley, S. et al. for the Heart and Estrogen/progestin Replacement Study Research Group. 1998. Randomized trial of estrogen plus progestin for secondary prevention of coronary heart disease in postmenopausal women. *Journal of the American Medical Association* 280:605–13.

Manson, J. E. et al., the Women's Health Initiative Investigators. 2003. Estrogen plus progestin and the risk of coronary heart disease. *New England Journal of Medicine* 349:523–34.

Pilon, D., A. Castilloux, and J. LeLorier. 2001. Estrogen replacement therapy: Determination of persistence with treatment. *Obstetrics & Gynecology* 97:97–100.

Rapp, Stephen R. et al. 2003. Effect of estrogen plus progestin on global cognitive function in postmenopausal women: The Women's Health Initiative Memory Study: A randomized controlled trial. *Journal of the American Medical Association* 289(20):2663–72.

Shumaker, S. A. 1998. The Women's Health Initiative Memory Study (WHIMS): A trial of the effect of estrogen therapy in preventing and slowing the progression of dementia. *Controlled Clinical Trials* 19:604–21.

Shumaker, S. A. et al. 2003. Estrogen plus progestin and the incidence of dementia and mild cognitive impairment in postmenopausal women: The Women's Health Initiative Memory Study: A randomized controlled trial. *Journal of the American Medical Association* 289:2651–62.

Shumaker, S. A. et al. 2004. Conjugated equine estrogens and incidence of probables dementia and mild cognitive impairment in postmenopausal women, Women's Health Initiative Memory Study. *Journal of the American Medical Association* 291:2947–58.

Stefanick, M. L. et al. 2003. The Women's Health Initiative postmenopausal hormone trials: Overview and baseline characteristics of participants. *Annals of Epidemiology* 13(9 Suppl): S78-86.

Stampfer, M. J. et al. 1991. Postmenopausal estrogen therapy and cardiovascular disease: Ten-year follow-up from the Nurses' Health Study [see comments]. *New England Journal of Medicine* 325:756–62.

Wassertheil-Smoller, Sylvia et al. 2003. Effect of estrogen plus progestin on stroke in postmenopausal women: The Women's Health Initiative: A randomized trial. *Journal of the American Medical Association* 289:2673–84.

The Women's Health Initiative Steering Committee. 2004. Effects of conjugated equine estrogen in postmenopausal women with hysterectomy: The Women's Health Initiative randomized controlled trial. *Journal of the American Medical Association* 291(14):1701–12.

Women's Health Initiative Study Group. 1998. Design of the Women's Health Initiative clinical trial and observational study. *Controlled Clinical Trials* 19:61–109.

Writing Group for the PEPI Trial. 1995. Effects of estrogen or estrogen/progestin regimens on heart disease risk factors in postmenopausal women: The Postmenopausal Estrogen/Progestin Interventions (PEPI) Trial. *Journal of the American Medical Association* 273:199–208.

Writing Group for the Women's Health Initiative Investigators. 2002. Risks and benefits of estrogen plus progestin in healthy postmenopausal women: Principal results from the Women's Health Initiative randomized controlled trial. *Journal of the American Medical Association* 288:321–33.

12

The Medicalization of Menopause in America, 1897–2000

Mapping the Terrain

Judith A. Houck

Menopause, the end of fertility in women, has long been associated in the United States with a bewildering array of symptoms. Hot flashes, memory loss, nervousness, vaginal dryness, night sweats, and inhibited sexual desire currently headline an ever-changing set of complaints. Throughout American history, most women have coped with their symptoms—should they occur—in a variety of ways: grinding their teeth, doting on grandchildren, drinking a shot of whiskey, buying a fan, running for office. In other words, most women have fought menopausal discomforts without calling on the ministrations of medicine. But in the last one hundred years or so, physicians have increasingly courted menopausal women as patients. Menopausal women, in turn, have sought medical care. As a result, during the twentieth century menopause became medicalized.

Feminists, in the academy and out, have recently decried the medicalization of menopause, claiming that physicians and drug companies have worked in tandem to transform menopause from a natural biological process into a pathological condition warranting medical intervention and pharmaceutical armament (Rostosky and Travis 1996; Kaufert and McKinlay 1985; Worcester and Whatley 1992; Klein and Dumble 1994;

Seaman 2003; Coney 1994). Indeed, for many American women reaching menopause early in the twenty-first century, menopause has become associated with medical appointments and prescription drugs in ways unforeseen by their grandmothers. This increased medical involvement has assuredly relieved a variety of symptoms, but it has also exposed women to health risks, the extent of which remain unknown.

Although menopause has undeniably come under increased medical scrutiny, the generalized critique of medicalization fails to adequately characterize the medical involvement with menopause at different moments in time. Medicalization is not an endpoint. Instead, it is a process that changes over time in response to cultural pressures and technological developments. Further, the discussion of medicalization too often credits medicine for the near-complete domination of the cultural meaning and personal experience of menopause. This overstates the reach of the biomedical model. In this chapter I attempt to describe the evolving role of medicine in the medical construction, popular understanding, and lived experience of menopause in the United States. I acknowledge the influence of medicine while simultaneously showing the limits of medicalization in the lives of menopausal women.

1897–1940: Physicians Notice Menopause

Before the end of the nineteenth century, U.S. physicians paid scant attention to menopause except to note, with no small degree of foreboding, that it was serious business. To the extent that it was discussed, menopause was described as a "critical period," a biological Rubicon that tested a woman's emotional and physical fortitude (Smith-Rosenberg 1986). However, physicians in the United States seemed largely unconcerned. Few American journal articles focused on menopause, and no American doctor published a monograph on menopause until Currier did in 1897. By then, physicians had begun to take menopause more seriously.

Paradoxically, when menopause began to attract medical attention, it seemed to have lost much of its medical significance. By the end of the nineteenth century, menopause no longer represented a biological Rubicon. Instead, the medical literature of the new century described it as an important milestone in a woman's life but denied that it was fraught with danger (Findley 1913; Lowry 1919; Strongin 1933). For example, one physician remarked that menopausal changes "come about as gently as the falling of the autumn leaves" (Cook 1903, 384). Others maintained that most women did not suffer at all during menopause (Drake 1902; Carr 1914; Hirst 1925).

Most physicians in the early twentieth century encouraged menopausal women to seek medical attention, but they explicitly denied that women needed medical treatment. Instead, physicians believed that information and reassurance were more valuable "than all the therapeutic agents under the sun" (Peple 1905, 644; see also Upshur 1905; Anspach 1924). As late as 1935, the eminent Johns Hopkins professor of gynecology Emil Novak maintained that physicians "earn their fees better through education and prevention than by writing out a prescription" (1935, 97–98). Physicians generally believed that an explanation and description of the physiological changes of menopause could alleviate most women's suffering. As one New York City gynecologist noted in 1933, if menopause were "properly understood by the woman herself . . . anxiety, together with actual dangers, might be largely eliminated" (Strongin 1933, 522).

Physicians coupled reassurance with a prescription for healthy habits, urging their menopausal patients to get lots of rest and daily exercise in the fresh air. A moderate diet, an avoidance of alcohol, and loose fitting clothing all promised to diminish the discomforts that emerged at middle age (Ashton 1907; Mosher 1918; Podolsky 1934). Notably, this prescription differed little from the regimen encouraged for all people, regardless of age, sex, or complaint.

Unlike physicians in the nineteenth century, however, doctors during this period had a new therapeutic tool to help ease a woman's journey through menopause—organotherapy, also referred to as ovarian therapy. The therapeutic promise of organotherapy was revealed in 1889 when the French-born physiologist Charles-Edouard Brown-Séquard injected himself with extracts from guinea pig and dog testes; he reported renewed vigor as a consequence (Oudshoorn 1994). By 1910, researchers in the United States had begun using ovarian preparations to treat menopausal symptoms, and by 1920, at least three pharmaceutical companies had manufactured ovarian extracts ("To Combat the Annoying Symptoms" 1910).

Despite the alleged promise of organotherapy, most physicians remained skeptical about its worth through the 1930s (even after estrogen and progesterone were isolated), claiming that ovarian therapy was crude, largely ineffective, and expensive. Even proponents of ovarian therapy conceded that most women did not need any pharmaceutical treatment (Hirst 1925; Swanberg 1937). For example, in 1922 Emil Novak argued that ovarian therapy could help relieve vasomotor symptoms, but he later insisted that most menopausal women "need no medical treatment whatsoever" (1922, 614–15; 1935, 97).

The increased interest in menopause expressed by U.S. physicians between 1897 and 1940 indicates a new level of medical involvement. Increased medical interest, however, does not reflect a medical takeover of menopause. Indeed, medical consensus during the period neither defined menopause as a medical problem nor claimed that menopause required the ministrations of medicine. Even the potential value of ovarian therapy failed to convince most physicians that menopause warranted medical intervention.

For their part, women seem to have agreed. Although some women surely felt discomfort at menopause, most failed to take those discomforts to physicians for relief (Peple 1905; Potter 1927). Perhaps loosening a corset, taking a swig of Dr. Pierce's Tonic, or participating in a suffrage parade provided sufficient relief for some (*Pierce's Memorandum Account Book* 1902; Mosher 1918). Even when menopausal women did seek a physician's counsel, they were more likely to receive information than the newest pharmaceutical weapon.

1941–1962: Medical Reluctance or a Therapeutic Revolution?

Before the 1940s, physicians, despite their increased interest, remained largely on the sidelines of the menopausal experience. Medicine became more involved, however, in the wake of two biomedical developments. First, in 1938, Sir Charles Dodds developed a synthetic hormone, diethylstilbestrol (DES). Although DES eventually gained notoriety for its use in preventing miscarriage, it was first prescribed for menopausal women after receiving approval from the Food and Drug Administration (FDA) in 1941 (Bell 1995; Seaman 2003). Second, in 1942, researchers at Ayerst, McKenna and Harrison, a Montreal-based drug company, developed an estrogen extract from the urine of pregnant mares. This product, marketed under the brand name Premarin, had all the benefits of DES with none of its potential side-effects. Taken together these developments reflect an important shift in the possible therapeutic regimen for menopause. But to what extent did the introduction of DES and Premarin medicalize menopause?

Medical Reluctance

Many elite physicians realized that these new forms of estrogen, which were readily available, inexpensive, and efficacious, represented a watershed in the options for medical treatment of menopause.[1] Rather than recommending their widespread use, however, physicians writing in the medical literature generally regarded hormones as a useful treatment for occasional cases. Indeed, physicians recommended a three-tiered treatment plan. First, they insisted that most menopausal symptoms could best be treated with reassurance and education. These physicians urged their colleagues to explain the physiology of menopause, discuss what women might experience, and emphasize that symptoms would eventually subside on their own (Emil Novak 1938; Creadick 1958; Owen 1945). Physicians believed that once women were armed with knowledge, even the hot flashes would "diminish in importance and cease to be troublesome" (Farquharson 1955, 202).

Despite their confidence in a reassuring word and a sympathetic ear to solve most women's menopausal difficulties, these physicians acknowledged that some women needed more than a heart-to-heart talk and a lesson in physiology. For women who needed something extra, many physicians recommended the short-term use of mild sedatives. In theory, sedatives helped women adjust to their new social circumstances or cope with their family crises (Creadick 1958; Freed 1950; Klingensmith 1954; Coleman 1947).

Only if reassurance or sedation failed did most physicians recommend hormone treatment for menopausal patients. Most doctors writing in the medical literature reserved hormonal therapy for women with severe, unrelenting symptoms, perhaps only 5 to 10 percent of women seeking medical attention (Danforth 1961; Hamblen 1940, for example). Most physicians agreed with Emil Novak, who claimed that in general "it seems better to let nature take its course except . . . when symptoms become very troublesome" (1945, 772).

Most physicians who did prescribe hormones between 1941 and 1962 argued strongly that treatment should be temporary, just long enough to help a woman adjust to her diminished estrogen levels (Payne 1952; Devereux 1947; Friedlander 1955; Klingensmith 1954). In the mid-1950s, however, a very few physicians began to promote indefinite use. The work of Fuller Albright and his colleagues at Harvard Medical School and Massachusetts General Hospital probably sparked this approach. In a 1941 article in the *Journal of the American Medical Association,* Albright and colleagues claimed that the "postmenopausal state" was the most significant factor in osteoporosis and that estrogen therapy helped bones to retain calcium (Albright, Smith, and Richardson 1941). Following the implications of their mentor's work, Albright's students Philip Henneman and Stanley Wallach believed that the long-term use of estrogens was important to halt the progression of postmenopausal osteoporosis (Henneman and Wallach 1957; Wallach and Henneman 1959). By 1953, other researchers were making similar claims about the ability of long-term estrogen therapy to reduce coronary atherosclerosis in postmenopausal women (Wuerst, Dry, and Edwards 1953; Robinson, Chen, and Higano 1958; Robinson, Higano, and Cohen 1959; and Rivin and Dimitroff 1959).

Building on the alleged illness-preventing benefits of estrogen replacement therapy, by the end of the 1950s, a handful of physicians started suggesting an aggressive interventionist role for the regimen. These physicians suggested that estrogen be given to all women at menopause to prevent the tragic consequences of "estrogen deprivation." In 1954, for example, Kost Shelton, clinical professor of medicine at the University of California, Los Angeles, claimed that menopause frequently transformed a woman into a "shell of the former alluring woman," and he advocated "estrogen substitution therapy" to allow women to remain "physiologically intact" throughout their lives (1954, 629). In the 1950s, Kost represented a notable exception to a more general conservative approach to estrogen therapy in the medical press. He also foreshadowed a development that emerged with more vigor in the 1960s.

Therapeutic Revolution

If we relied on the medical press exclusively to illuminate the medicalization of menopause, however, our picture would be significantly skewed. Indeed, during this period of medical conservatism women routinely began to use estrogen therapy to ease their menopausal symptoms.

Two surveys from the 1950s showed that roughly 30 to 35 percent of the highly educated middle-class white women surveyed received hormone treatments (Brush 1950; Davidoff and Platt 1957). How can we reconcile the conservative path recommended in the medical press with the widespread use of hormones?

Several factors contributed to prescription rates higher than those that might be predicted by the medical literature. First, the use of estrogens in this period must be understood within the context of a larger therapeutic revolution. The development of sulfa drugs and penicillin in the 1930s and 1940s created a public demand for the so-called miracle drugs, and patients increasingly expected physicians to end a medical encounter by prescribing a drug (Pellegrino 1979).

Second, this drug revolution was coupled with aggressive marketing strategies by the drug industry. After World War II, drug companies began advertising their products in earnest to physicians, both through advertisements in medical journals and visits from drug salesmen. Indeed, physicians in the 1950s were much more likely to learn about drug efficacy and advisability from a salesman than from a journal article (Silverman and Lee 1974). Consequently, many physicians found the new drugs were more attractive than sending their patients home with a pat on the knee or words of kindness and advice.

Patient demand provides a third and related explanation for the widespread use of hormones during this period. Evidence suggests that women approached their doctors to ask for hormone therapy. Indeed, physicians often expressed frustration at the unreasonable expectations that women had for estrogen. One physician complained, for example, that women "actually beg the physician" for hormones (Biskind 1957, 117; see also Rogers 1956; Galton 1950; Watson 1944; Hamblen 1949). The depictions of hormone replacement therapy in popular magazines encouraged these women to seek medical attention.

While the medical press promoted caution, the popular press, sometimes prodded by the drug companies, enthusiastically touted the wonders of estrogen therapy. Magazines and popular books heralded hormone treatments as "one of the most brilliant accomplishments of modern medicine" (Loomis 1939, 232). Claiming that the march of medical progress was nothing short of divine, Albert Maisel promoted in *Woman's Home Companion* the "medical miracles" brought by hormone therapy (1954, 41). Other health writers goaded their readers into seeking out hormone therapy, dismissing women's resistance as wrong-headed and old-fashioned (Bailey 1947).

Popular characterizations made it seem reasonable and progressive for women to seek medical relief at menopause. The modern woman should reject the plight of her grandmother, resist the role of martyr, and take control of her body and her health. Medical science made that control possible. Even when these authors claimed that most women would not need hormones, women with symptoms might have assumed that *their* situation required medical intervention.

Despite the conservative approach to hormones promoted in the medical literature, the enthusiastic accounts in popular magazines and the promotional efforts of drug companies raised the profile of hormones between 1941 and 1962. As a result, many middle-class women began to see hormones (and sedatives) as appropriate responses to the trials of menopause. Nevertheless, medicalization was far from absolute during this period. The limited evidence suggests that of the women most likely to seek medical care during this period—educated, upper- and middle-class white women—no more than 30 percent used hormones. For women outside this demographic, medical involvement was probably significantly less frequent. As a result, for most women during this period, menopause remained a process untouched by medical intervention.

1962–1975: Menopause as an Estrogen Deficiency Disease

In 1962, the Johns Hopkins obstetrician-gynecologist Allan Barnes posed a question to the readers of *Consultant:* "Is Menopause a Disease?" He answered affirmatively, listing the consequences of estrogen deficiency and concluding, "menopause is a disease process, requiring active intervention" (23). Other physicians pledged their support for this idea. Frances Rhoades, a Detroit physician specializing in geriatrics, characterized menopause as a "chronic and incapacitating deficiency disease" and insisted that it was both "morally and medically justifiable" to provide long-term hormone treatment (1965, 410, 412). The language of disease, so jarring in these articles, marks a new development in the medical involvement with menopause.

The "disease model" of menopause claimed that menopause could be accurately (and fruitfully) understood as an estrogen deficiency disease. With this frame, menopause became comparable to conditions more commonly understood as diseases such as diabetes. Indeed, just as insulin deficiency produced diabetes, so too did estrogen deficiency produce menopause. In both cases, the "cure" was clear: provide the missing element.

It is tricky to date precisely the emergence of this general understanding of the menopausal transition. Surely, throughout most of the twentieth century, some physicians had recommended estrogen (or its precursor, ovarian extracts) to compensate for the natural depletion of hormones. It could be argued, then, that the disease model emerged at the same time as hormonal therapy. But four new characterizations of menopause and its treatment came together only in the early 1960s and suggest a significantly new understanding of menopause.

First, the creation of "estrogen deficiency disease" blamed menopause for a wide range of conditions more commonly understood as consequences of aging. Wrinkles, osteoporosis, heart disease, depression, sexual dysfunction, dull hair, sagging breasts, hypertension, and alcoholism were all alleged to be the tragic consequences of estrogen deficiency under the new disease model. Although physicians had previously linked increased rates of heart disease and osteoporosis to decreased postmenopausal estrogen levels, these earlier physicians had not indicted menopause itself as the censurable culprit.

Second, the disease model often hinged on the claim that menopausal women, by outliving their reproductive potential, were themselves "unnatural." Arguing that menopausal women were "evolutionary accidents" or products of modern medicine, which had unnaturally extended the female lifespan, estrogen deficiency proponents legitimated aggressive medical intervention (Kupperman 1972; Kistner 1973). Before the disease model took hold, some physicians touted menopause as evidence of evolutionary wisdom for allowing women to raise, rather than only to bear, her children.

Third, the disease model proposed giving replacement hormones to women, regardless of whether they exhibited symptoms. The problem, according to the model, was not the hot flash or nervous anxiety; the problem was estrogen deficiency. As a result, adherents to this construction argued that withholding hormones from women just because they complained of no discomfort flew in the face of responsible preventative medicine (Davis, Strandjord, and Lanzl 1966; Greenblatt 1974).

Finally, the disease model often explicitly used the language of disease. Certainly, descriptions of menopause that referred to "estrogen deprivation," which "required" medical treatment (Shelton 1954), suggest that early formulations of the disease model did not necessarily depend on the use of the particular word *disease.* Nevertheless, the rising concern about the perils of estrogen deficiency generally coincided with the rhetorical application of the word *disease* (Rhoades 1965; Davis 1967). The doctors who promoted this model self-consciously presented their conception of menopause as both new and groundbreaking.

The most famous proponent of the disease model of menopause was a Brooklyn gynecologist, Robert A. Wilson. Although he first published his controversial views in the medical press (Wilson and Wilson 1963; Wilson 1962; Wilson, Brevetti, and Wilson 1963; Wilson 1964), he achieved fame of a sort by taking his message of despair and hope to the general public in a 1966 book called *Feminine Forever. Feminine Forever* and its medically oriented precursors, detailed the consequences of "Nature's defeminization." Wilson claimed that estrogen depletion led to hypertension, high cholesterol, osteoporosis, and arthritis. In addition, he and his collaborator (and wife) Thelma A. Wilson insisted that menopause frequently led to serious emotional disturbances; even women who escaped debilitating depression frequently acquired a "vapid cow-like feeling called a negative state" (1963, 352). They knowingly maintained that these women "see the world through a gray veil, and they live as docile creatures missing life's values" (353). Indeed, the Wilsons believed that "these women exist rather than live" (351).

Robert Wilson did not, however, abandon menopausal women to their dreary fate. Rather, he promised women a pharmaceutical escape route—estrogen replacement therapy (ERT). Comparing ERT to insulin, Wilson insisted that replacement therapy could both cure and prevent estrogen deficiency disease. By allowing women to remain "fully sexed," long-term hormone therapy prevented the "supreme tragedy" of women's lives (Wilson 1966, 18, 105).

The popular media swiftly publicized Wilson's claims. By 1964, *Time* and *Newsweek* had published articles extolling the promise of hormone therapy to cure menopause, and women's magazines followed in 1965 ("No More Menopause" 1964; "Durable, Unendurable Women" 1964; Kaufman 1965). The publicity intensified after the publication of *Feminine Forever.* The book itself sold more than 100,000 copies in the first seven months after its release (Mintz 1969), and it was serialized widely in local newspapers. Perhaps most important, it generated an interest in menopause that was reflected in a flood of popular articles and books.

Wilson's views, though echoing some others in the profession, did not represent medical orthodoxy, although long-term estrogen therapy did gain support during this period. Instead, physicians fell into one of four loosely defined camps. A small group of cautious physicians refused to use hormones under any circumstances. Another group prescribed hormone therapy for women with severe symptoms but denounced long-term treatment (Hertz 1967; Kase 1974; Edmund Novak, Jones, and Jones 1975). Most of these physicians reserved hormone therapy for addressing the symptoms most clearly linked to diminished estrogen—vaginal atrophy and hot flashes. A third group did not advocate hormone therapy for all menopausal women but recommended long-term estrogen therapy as needed. Although these physicians supported long-term therapy, they scrambled to distance themselves from the unreasonable expectations associated with it. They insisted that not all women required hormone treatments but maintained that ERT provided miraculous relief for those women who experienced menopausal difficulties (Israel 1961; Nachtigall 1977; Kistner and Utian 1979). A fourth school of thought generally considered menopause a deficiency disease and therefore recommended hormone replacement at the first sign of estrogen depletion, regardless of whether women exhibited any symptoms. Despite being quite radical, this group included several important figures, including M. E. Davis (1967), a University of Chicago gynecologist, and Allan C. Barnes (1962), a gynecologist at Johns Hopkins, for example. Between 1962 and 1975, no one position could claim to represent medical orthodoxy.

The popular messages about menopause and its treatment reflected the medical ambivalence. In general, the popular literature portrayed menopause as a potential problem for women and promoted estrogen replacement therapy as its solution. Until 1975, virtually all the articles depicted replacement hormones positively—at least in the short term—focusing on the "personal miracles" that ERT performed (Rorvik 1971, 102). In addition to assuring women that hormones would relieve hot flashes and vaginal dryness, one author also claimed that it was "well-established" that hormones could prevent heart disease (Rice 1975,128b).

Occasionally, writers prodded women into seeking out the modern solution for menopausal troubles. In a 1967 *Good Housekeeping* article, a doctor compared estrogen treatment with glasses and hair color. "If you couldn't read the fine print, you'd get glasses. . . . And you've probably tried rinses for your hair. Take female hormones in the same way—a restoration of what used to be" ("Change of Life" 1967, 19–20). Another medical writer presented ERT as the modern choice for the modern woman. He claimed that "these are changing times: medical science has made such advances that change of life can be looked at in a change of light" ("The Menopause That Refreshes" 1973, 87).

Significantly, however, the popular literature did not generally maintain that all menopausal women would benefit from hormone replacement. Instead, it emphasized that most women did not need estrogen therapy or any other medical treatment. One article, for example, noted that only 25 percent of postmenopausal women "have hormone deficiencies." The author suggested that for most women, hormone therapy was "downright silly" (Naismith 1966, 101). Although these articles did indeed "promote" estrogen therapy, they did not recommend long-term (or even short-term) treatment for all menopausal women.

The popular literature also provided conflicting ideas about whether menopause was a disease. On the one hand, women could find support in popular magazines that menopause was a "deficiency state which gradually defeminizes a woman physically" (Kaufman 1965, 23). On the other hand, some articles challenged this assumption head-on by insisting that menopause was a natural process that needed no "cure" (Cherry 1976).

Given the disagreements within the information available to women, how can we understand the significance of this literature? The popular literature did indeed publicize and promote ERT as a wondrous cure for the negative consequences of menopause. Certainly, Wilson's work and some other books and articles encouraged women to seek out long-term hormone therapy by dangling before them the consequences of untreated menopause and the modern miracles of hormone treatment. Although women could and probably did encounter this position while they thumbed through women's magazines, it did not represent the only view of menopause available during this period. In contrast to Robert Wilson's outrageous rhetoric, the popular literature did not overwhelmingly present ERT as a necessary or desirable treatment for all women approaching middle age. Instead, it presented ERT as a valuable treatment for women experiencing a particularly difficult menopause. The wide-ranging opinions then allowed women to choose among several competing meanings of menopause and treatment regimens.

Despite the lack of medical consensus, the use of estrogen therapy did indeed increase after 1962. Drug companies followed the free publicity gained by Robert Wilson's efforts with their own advertising campaign aimed at informing physicians of the wide-ranging benefits of hormone therapy (Watkins 2001). Indeed, Robert Wilson's research foundation, devoted to the elimination of "estrogen deficiency disease," was funded in part by donations from Searle, Ayerst, and Upjohn. Further, some women, encouraged by the promise of relief from hot flashes or compelled by the image of a more youthful appearance, demanded hormones from their physicians (Houck 2003). In time, most physicians acquiesced to pressure from both sides and from the desire to help their menopausal patients. Some doctors eventually saw hormone replacement therapy as a godsend for menopausal women, providing both long-term and short-term benefits with no apparent drawbacks. Other physicians prescribed hormones more reluctantly, giving in to their patients' demands while still believing that most women needed only an encouraging talk and a stiff upper lip to see them through the rough times. Regardless of the lingering doubts of some physicians, dollar sales of noncontraceptive estrogen more than quadrupled between 1962 and 1975 (Greenwald, Caputo, and Wolfgang 1977).

As startling as these numbers are, interpreting them against the backdrop of Robert Wilson's campaign is tricky. It is tempting, of course, to attribute much of the increase to the efforts of Robert Wilson and likeminded physicians. Perhaps this is partially true, but the full explanation is more complex. The use of prescription drugs in most categories, for example, doubled during this period (Waldron 1977). Further, the price of estrogen itself nearly doubled between 1962 and 1975 (IMS America 1962, 1966, 1976). Taken together, these two facts suggest that even without the work of Robert Wilson, the dollar sales of estrogens would have increased significantly. Regardless of the cause, however, between 1965 and 1974, the number of women receiving their first prescription for estrogen replacement therapy roughly doubled (IMS America 1975). In sheer numbers alone, then, medicine became a more prominent presence in the menopausal experience.

Clearly, the medicalization of menopause continued and expanded between 1962 and 1975. But, again, it is easy to overstate the reach of this medical involvement. Although some physicians argued that menopause was a deficiency disease, most did not. Although some physicians urged the long-term use of hormones, most did not. Although more women than ever sought and received medical advice and treatment for menopause, most did not. As a result, the medicalization of menopause did not rob patients or physicians of all their options; medicalization was hardly absolute.

1975 and Beyond: The Medicalization of Aging

In 1975, at the height of estrogen's popularity, two articles in the *New England Journal of Medicine* dealt the therapy a significant blow. Researchers at Washington University and Kaiser-Permanente Medical Center independently discovered a link between postmenopausal estrogen therapy and endometrial cancer (Ziel and Finkle 1975; Smith et al. 1975). Although researchers had proposed a link between estrogen and cancer since the 1940s, these landmark studies supplied the best evidence at the time that ERT posed a cancer risk in humans.

The risk of endometrial cancer did not lead physicians to abandon hormone replacement therapy altogether after 1975. Instead, they proceeded more cautiously. Physicians wrote more than twenty-eight million prescriptions for replacement hormones in 1975; in 1980, they wrote only fifteen million (Worcester and Whatley 1992). Furthermore, they generally lowered the dosage and shortened the duration of treatment (Pasley, Standfast, and Katz 1984).

The setback for hormones was temporary, however. Hormone therapy rebounded in the early 1980s; the number of prescriptions for replacement estrogens rose from 13.6 million in 1982 to 31.7 million in 1992 (Watkins 2001). By the 1990s, something like a medical consensus emerged around the claim that long-term hormone therapy was good medicine (Grady and Rubin 1993; "Hormone Replacement Therapy" 1993). In 1998, one survey found that 34 percent of women aged fifty or older used HRT (up from 23 percent in 1993; the figures are from Collins et al. 1999).

Several factors, both medical and social, help explain the recovery. First, the aging baby-boomer cohort had swelled the ranks of women facing menopause. Between 1970 and 2000, the number of women aged forty-five to fifty-four increased by roughly 56 percent (or more than 6.5 million women). Second, these women continued to want relief from hot flashes, vaginal dryness, and other discomforts (be they necessarily linked to menopause or not). Hormones effectively relieve at least some meno-pausal symptoms. Third, physicians routinely added progestin to the es-trogen treatment, a regimen that significantly reduced the risk of en-dometrial cancer. While this combination had been prescribed for years, it became the standard hormonal therapy for women with a uterus only in the 1980s. Fourth, the prevention of osteoporosis and heart disease emerged as the primary indications for sustained hormone use. Although some medical researchers had championed estrogens to prevent these diseases for decades, only in the late 1970s (for osteoporosis) and the 1990s (for heart disease) did they figure prominently in the marketing of hormone replacement therapy (Watkins 2001).

What does the fall and rise of hormone replacement therapy tell us about the medicalization of menopause? Surely, in the 1990s, more Amer-ican women received a prescription for replacement hormones at meno-pause than ever before. Further, women increasingly used this therapy to prevent illness in addition to relieving menopausal symptoms. This latter development, however, complicates our understanding of the medicaliza-tion of menopause at the close of the twentieth century. The treatment that once helped gauge the level of medical involvement in menopause is now offered as a generalized therapy for the consequences of female ag-ing (Watkins 2001). Perhaps, then, the medicalization of menopause did not intensify in the 1980s and 1990s; instead, menopause provided a gate-way for the medicalization of female middle age.

But, again, let us not lose sight of the roughly 66 percent of American women older than fifty who, despite the hard sell by drug companies and the efforts of well-intentioned physicians, nevertheless failed to embrace medical intervention (Collins et al. 1999). Certainly, some of these women may have wanted hormones but limited access to health care thwarted their desires. But other women, perhaps influenced by the feminist cri-tique of medicalization or encouraged by a willingness to seek alternative healing paths, intentionally looked elsewhere for ways to cope. Some women found symptom relief in soy products and black cohosh. Others renamed hot flashes "power surges" and embraced them as a source of feminine inspiration. Still others gritted their teeth or bought a fan. And, finally, some women barely noticed menopausal symptoms. At the apex of hormone replacement use, then, menopausal women continued to seek out and employ alternatives to medical recommendations.

Recent studies (as detailed elsewhere in this volume) raise questions about the future of hormone replacement therapy and the medicalization of menopause. But as we continue to track and analyze medical involvement in the lives of menopausal women, we must also gauge the reach of medical influence. Despite the increased medicalization of countless behaviors and bodily functions (Zola 1990) during the twentieth century, most menopausal women in the United States have not succumbed, at least not entirely, to medical dictates.

Note

1. Most of the physicians publishing in the medical literature held university appointments

References

Albright, Fuller, Patricia H. Smith, and Anna M. Richardson. 1941. Postmenopausal osteoporosis: Its clinical features. *Journal of the American Medical Association* 116:2465–74.

Anspach, Brooke M. 1924. *Gynecology.* 2d ed. Philadelphia: J. B. Lippincott.

Ashton, William Easterly. 1907. *A text-book on the practice of gynecology for practitioners and students.* 3d ed. Philadelphia: W. B. Saunders.

Bailey, Bernadine. 1947. Fair, fit, and forty. *Hygeia,* December, pp. 930–31.

Barnes, Allan C. 1962. Is menopause a disease? *Consultant* 2:22–24.

Bell, Susan E. 1995. Gendered Medical Science. *Feminist Studies* 21:469–500.

Biskind, Leonard. 1957. *Health and hygiene for the modern woman.* New York: Harper.

Brush, Dorothy. ca. 1950. Papers from the Sophia Smith Collection and College Archives. Smith College, Northampton, Mass.

Carr, C. S. 1914. The menopause or change of life. *Physical Culture* 32:521–25.

Change of life. 1967. *Good Housekeeping,* September, pp. 10, 12, 16, 20, 22.

Cherry, Sheldon. 1976. *The menopause myth.* New York: Ballantine.

Coleman, Foster. 1947. The menopause. *Kentucky Medical Journal* 45:207–9.

Collins, Karen Scott et al. 1999. *Health concerns across a woman's lifespan: The Commonwealth Fund 1998 survey of women's health.* New York: Commonwealth Fund.

Coney, Sandra. 1994. *The menopause industry: How the medical establishment exploits women.* Alameda, Calif.: Hunter House.

Cook, George Wythe. 1903. Some observations on the menopause. *American Journal of Obstetrics* 45:382–86.

Creadick, Robert N. 1958. Menopause. *Texas State Journal of Medicine* 54:709–11.

Currier, Andrew Fay. 1897. *The menopause: A consideration of the phenomena which occur to women at the close of the child bearing period.* New York: Appleton.

Danforth, David N. 1961. The climacteric. *Medical clinics of north america* 45:47–52.

Davidoff, Ida, and Marjorie Platt. 1957. Two generation study of postparental women. Henry A. Murray Research Center, Radcliffe College, Cambridge, Mass.

Davis, M. E. 1967. The physiology and management of the menopause, pp. 419–31. In *Advances in obstetrics and gynecology,* edited by Stewart L. Marcus and Cyril C. Marcus. Baltimore: Williams and Wilkins.

Davis, M. E., Nels M. Strandjord, and Lawrence Lanzl. 1966. Estrogens and the aging process: The Detection, prevention, and retardation of osteoporosis. *Journal of the American Medical Association* 196 (3): 219–24.

Devereux, William. 1947. Management of menopause and climacteric. *Texas State Journal of Medicine* 42:683–87.

Drake, Emma F. 1902. *What a woman of forty-five ought to know.* Philadelphia: Vir Publishing.

Durable, unendurable women. 1964. *Time,* October 16, p. 72.

Farquharson, R. F. 1955. The menopausal patient. *Medical Record and Annals* 49:196–204.

Findley, Palmer. 1913. *A treatise on the diseases of women for students and prac-titioners.* New York: Lea and Febiger.

Freed, S. Charles. 1950. The menopausal syndrome. *Journal of Insurance Medicine* 5:21–25.

Friedlander, Harry. 1955. Education and sedation in menopausal therapy. *Post-graduate Medicine* 18:94–98.

Galton, Lawrence. 1950. What every husband should know about the menopause. *Better Homes and Gardens,* July, pp. 54–55.

Grady, Susan, and Susan Rubin. 1993. Postmenopausal hormone therapy. *Annals of Internal Medicine* 119:347–48.

Greenblatt, Robert B. 1974. Reprise, pp. 222–31. In *The menopausal syndrome,* edited by Robert B. Greenblatt, Virendra B. Mahesh, and Paul G. McDo-nough. New York: Medcom Press.

Greenwald, Peter, Thomas A. Caputo, and P. E. Wolfgang. 1977. Endometrial cancer after menopausal use of estrogens. *Obstetrics and Gynecology* 50:239–43.

Hamblen, Edwin Crowell. 1940. The female climacteric. *Virginia Medical Monthly* 67:24–29.

———. 1949. *Facts About the Change of Life.* Springfield, Ill.: C. C. Thomas.

Henneman, Philip H., and Stanley Wallach. 1957. The use of androgens and es-trogens and their metabolic effects: A review of the prolonged use of estro-gens and androgens in postmenopausal and senile osteoporosis. *AMA Archives of Internal Medicine* 100:715–23.

Hertz, Roy. 1967. The role of steroid hormones in the etiology and pathogenesis of cancer. *American Journal of Obstetrics and Gynecology* 98:1013–19.

Hirst, John Cooke. 1925. *A manual of gynecology,* 2d ed. Philadelphia: W. B. Saunders.

Hormone replacement therapy. 1993. *International Journal of Gynecology and Obstetrics* 41:194–202.

Houck, Judith. 2003. "What do these women want?": Feminist responses to *Feminine forever,* 1963–1980. *Bulletin of the History of Medicine* 77:103–32.

IMS America. 1962. *National prescription audit.* Ambler, Pa.: IMS Research Group.

———. 1966. *National prescription audit.* Ambler, Pa.: IMS Research Group.

———. 1975. *National prescription audit: Ten-year trend.* Ambler, Pa.: IMS Research Group.

———. 1976. *National prescription audit.* Ambler, Pa.: IMS Research Group.

Israel, S. Leon. 1961. The menopause. *Postgraduate Medicine* 30:420–21.

Kase, Nathan. 1974. Estrogens and the menopause. *Journal of the American Medical Association* 227:318–19.

Kaufert, Patricia A., and Sonja M. McKinlay. 1985. Estrogen replacement therapy: The production of medical knowledge and the emergence of policy, pp. 113–38. In *Women, health, and healing: Toward a new perspective,* edited by Ellen Lewin and Virginia Olesen. New York: Tavistock.

Kaufman, Sherwin A. 1965. The truth about female hormones. *Ladies' Home Journal* January, pp. 22–23.

Kistner, Robert. 1973. The menopause. *Clinical Obstetrics and Gynecology* 16:106–29.

Kistner, Robert, and Wulf H. Utian. 1979. Estrogen replacement in the menopause. *Obstetrics and Gynecology Annual* 8:386–87.

Klein, Renate, and Lynette J. Dumble. 1994. Disempowering midlife women: The science and politics of hormone replacement therapy (HRT). *Women's Studies International Forum* 17:327–43.

Klingensmith, Paul O. 1954. The care of the menopausal patient. *Pennsylvania Medical Journal* 57:426–28.

Kupperman, Herbert. 1972. The climacteric syndrome. *Medical Folio* 15:1–2.

Loomis, Frederic. 1939. *Consultation room.* New York: Alfred A. Knopf.

Lowry, Edith Belle. 1919. *The woman of forty.* Chicago: Forbes.

Maisel, Albert Q. 1954. Promise for happiness. *Woman's Home Companion,* September, p. 41.

The menopause that refreshes. 1973. *Harper's Bazaar,* August, p. 87.

Mintz, Morton. 1969. *The pill: An alarming report.* Greenwich, Conn: Fawcett.

Mosher, Clelia Duel. 1918. *Health and the woman movement.* New York: Woman's Press.

Nachtigall, Lila. 1977. *The Lila Nachtigall report.* New York: G. P. Putnam.

Naismith, Grace. 1966. Common sense and the "femininity pill." *Reader's Digest,* September, pp. 99–102.

No more menopause? 1964. *Newsweek,* January 13, p. 53.

Novak, Edmund R., Georgeanna Seegar Jones, and Howard Jones. 1975. *Gynecology.* Baltimore: Williams and Wilkins.

Novak, Emil. 1922. An appraisal of ovarian therapy. *Endocrinology* 6:599–620.

———. 1935. *The woman asks the doctor.* Baltimore: Williams and Wilkins.

———. 1938. The menopause and its management. *Journal of the American Medical Association* 110: 619–22.

———. 1945. Some misconceptions and abuses in gynecological organotherapy. *Pennsylvania Medical Journal* 48:771–74.

Oudshoorn, Nelly. 1994. *Beyond the natural body: An archeology of sex hormones.* New York: Routledge.

Owen, Trevor. 1945. The medical view of the menopause. *American Journal of Psychiatry* 101:756–59.

Pasley, Beverly H., Susan Standfast, and Selig Katz. 1984. Prescribing estrogen during menopause: Physician survey of practices in 1974 and 1981. *Public Health Reports* 99:424–29.

Payne, Franklin L. 1952. The postmenopausal patient. *Journal of the Medical Association of the State of Alabama* 22:31–36.

Pellegrino, Edmund D. 1979. The sociocultural impact of twentieth-century therapeutics, pp. 245–66. In *The therapeutic revolution: Essays in the social history of American medicine,* edited by Morris Vogel and Charles Rosenberg. Philadelphia: University of Pennsylvania Press.

Peple, W. Londes. 1905. The menopause. *Carolina Medical Journal* 53:640–44.

Pierce's memorandum account book: Designed for farmers, mechanics, and all people. 1902. Buffalo, N.Y.: World's Dispensary Medical Association.

Podolsky, Edward. 1934. *Young woman past forty: A modern sex and health primer of the critical years.* New York: National Library Press.

Potter, Marion Craig. 1927. A new standard of health for the menopause. *Medical Woman's Journal* 34 (June): 157–60.

Rhoades, Frances. 1965. The menopause: A deficiency disease. *Michigan Medicine* 64:410–12.

Rice, Dabney. 1975. Anti-aging pill. *Harper's Bazaar,* August, p. 78.

Rivin, A. U., and S. P. Dimitroff. 1954. Incidence and severity of atherosclerosis in estrogen treated males and in females with hypoestrogenic or hyperestrogenic state. *Circulation* 9:533–39.

Robinson, R. W., W. D. Chen, and N. Higano. 1958. Estrogen replacement therapy in women with coronary atherosclerosis. *Annals of Internal Medicine* 48:95–101.

Robinson, R. W., N. Higano, and W. D. Cohen. 1959. Increased Incidence of coronary heart disease in women castrated prior to menopause. *AMA Archives of Internal Medicine* 104:908–13.

Rogers, Joseph. 1956. The menopause. *New England Journal of Medicine* 254:697–704, 750–56.

Rorvik, David M. 1971. You can stop worrying about the menopause. *McCall's,* October, pp. 102–3.

Rostosky, Sharon Scales, and Cheryl Brown Travis. 1996. Menopause research and the dominance of the biomedical model, 1984–1994. *Psychology of Women Quarterly* 20:285–312.

Seaman, Barbara. 2003. *The greatest experiment ever performed on women: Exploding the estrogen myth.* New York: Hyperion.

Shelton, E. Kost. 1954. The use of estrogen after the menopause. *Journal of the American Geriatrics Society* 2:627–33.

Silverman, Milton, and Philip R. Lee. 1974. *Pills, profits, and politics.* Berkeley: University of California Press.

Smith, Donald C., Ross Prentice, Donovan J. Thompson, and Walter L. Herrman. 1975. Association of exogenous estrogen and endometrial carcinoma. *New England Journal of Medicine* 293:1164–67.

Smith-Rosenberg, Carroll. 1986. Puberty to menopause: The cycle of femininity in nineteenth-century America. *Disorderly Conduct: Visions of Gender in Victorian America.* New York: Oxford University Press, pp. 182–96.

Strongin, Herman F. 1933. Woman—Her critical decade. *American Medicine* (November): 532–36.

Swanberg, Harold. 1937. The control of menopausal symptoms. *Illinois Medical Journal* 72:441–45.

To combat the annoying symptoms of the menopause. 1910. *New York Medical Journal* 91:756.

Upshur, John Nottingham. 1905. The menopause. *New York Medical Journal* 82:650–53.

Waldron, Ingrid. 1977. Increased prescribing of valium, librium, and other drugs—An example of the influence of economic and social factors on the practice of medicine. *International Journal of Health Services* 7:37–62.

Wallach, Stanley, and Philip H. Henneman. 1959. Prolonged estrogen therapy in postmenopausal women. *Journal of the American Medical Association* 171:1637–42.

Watkins, Elizabeth Siegel. 2001. Dispensing with aging: Changing rationales for long-term hormone replacement therapy, 1960–2000. *Pharmacy in History* 43:23–37.

Watson, B. P. 1944. The menopausal patient. *Journal of Clinical Endocrinology* 4:571–74.

Wilson, Robert A. 1962. The roles of estrogen and progesterone in breast and genital cancer. *Journal of the American Medical Association* 182:327–33.

———. 1964. The estrogen cancer myth. *Clinical Medicine* 71:1343–52.

——— 1966. *Feminine forever.* New York: M. Evans.

Wilson, Robert A., and Thelma A. Wilson. 1963. The fate of the nontreated postmenopausal woman: A plea for the maintenance of adequate estrogen from puberty to the grave. *Journal of the American Geriatrics Society* 11:347–62.

Wilson, Robert A., Raimondo E. Brevetti, and Thelma A. Wilson. 1963. Specific procedures for the elimination of the menopause. *Western Journal of Surgery, Obstetrics & Gynecology* 71:110–21.

Worcester, Nancy, and Marianne H. Whatley. 1992. The selling of HRT: Playing on the fear factor. *Feminist Review* 41:1–26.

Wuerst, J. H. Jr., T. J. Dry, and J. E. Edwards. 1953. Degree of Coronary atherosclerosis in bilaterally oophorectomized women. *Circulation* 7:801–9.

Ziel, Harry K., and William D. Finkle. 1975. Increased risk of endometrial carci-
    noma among users of conjugated estrogens. *New England Journal of Medi-
    cine* 293:1167–70.
Zola, Irving Kenneth. 1990. Medicine as an institution of social control, pp. 398–
    408. In *The sociology of health and illness: critical perspectives,* edited by
    Peter Conrad and Rochelle Kern. 3d ed. New York: St. Martin's Press.

13

The History of Hormone Replacement Therapy:

A Timeline

Barbara Seaman

I am a science journalist who has followed the "hormone story" for about forty-five years. I have watched the rise and fall and resurrection of various drugs, and seen patients' lives both helped and hurt by them. In the pages that follow, I try to provide an outline of many crucial events in the extended history of hormone replacement therapy (HRT). As much information in the timeline indicates, many scientists were aware of the dangers of the medicinal use of hormones well before the July 2002 decision to halt the estrogen plus progestin arm of the Women's Health Initiative study.

I would ask the reader to wonder why it took more than half a century from initial research on the dangers of hormone use to the widespread recognition of the dangers of HRT. Could it be that some significant causes of this lag are the blindness created by drug industry profits, sexism, patients' wishful thinking, and/or professional arrogance?

1890s

- Merck manufactures Ovariin, a coarse brownish powder derived from the pulverized ovaries of the cow, for the treatment of menopause. It remains on the market until 1932.

1922

- The anatomist Edgar Allen describes the hormone cycle in a female mouse. With the biochemist Edward Doisy, he develops the Allen-Doisy test to measure estrogen content.

Early 1930s

- Menopause products are derived from human pregnancy urine by the biochemists Adolph Butenandt, financed by Schering in Germany, and James Bertram Collip, financed by Ayerst Laboratories in Canada.
- In 1932, Antoine Lacassagne of the Institute Pasteur administers Butenandt's estrogen (Folliculin) to mice and induces mammary cancer.
- Due to the instability of the dosages, as well as the odor, by the late 1930s these formulations are succeeded by products from the urine of pregnant mares. The Canadian version, Premarin (or PREgnant MAres' uRINe), will become the most popular menopause treatment in the United States.

1937

- Schering patents ethinyl estradiol, now the most widely used contraceptive estrogen. It is also used in menopause treatments such as Pfizer's Femhrt.

1938

- In an effort to stop Hitler from cornering world estrogen markets, Edward Charles Dodds, a London biochemist and consulting physician to King George, publishes his formula for diethylstilbestrol (known as stilbestrol, and later as DES), a synthetic estrogen, in *Nature* magazine. In the months after Dodds's formula is published—free to copy by all comers, effective by mouth, and inexpensive—dozens of drug companies worldwide begin to manufacture it. Dodds soon becomes alarmed at the unproven uses—the drug is prescribed for everything from preventing miscarriage to maintaining youth—and begins a personal crusade to raise awareness of his formula's dangers. He notes: "Within a few months of the first publication of the synthesis of stilbestrol, the substance was being marketed throughout the world. No long term toxicity tests on animals were done first."

- The *Journal of the American Medical Association (JAMA)* and *Endocrinology* publish influential reports on the use of estrogen in menopause, confirming the benefits of the hormone in curbing hot flashes but also warning that long-term use might be carcinogenic. Neither paper suggests that estrogen could delay aging in any way.

1940

- In May, Eli Lilly asks the U.S. Food and Drug Administration (FDA) for permission to market diethylstilbestrol. Lilly's application papers include a report by Drs. K. K. Chen and P. N. Harris listing studies in mice and rats that show that males as well as females may get breast cancer when administered diethylstilbestrol. Inter-sexuality appears in male rats, mice, and chickens; milk production in animals is reduced, and abortion in cows is observed.
- In November, Lilly withdraws its application after the FDA advises that it will be rejected. Ten other companies, including Squibb, Upjohn, Abbott Laboratories, and Sharp & Dohme withdraw their applications for the same reason.

1941

- The companies that wish to market diethylstilbestrol meet in Washington, D.C. What follows is the first blueprint for joint spin-doctoring by the drug companies, and it provides the basis for Big Pharma (Pharmaceutical Research and Manufacturer of America). The group hires Carson P. Frailey, executive vice president of the American Drug Manufacturers Association, to guide them in a strategy. He advises them to enlist doctors from around the country in a letter-writing campaign to the FDA. Many doctors cooperate, and under tight discipline the companies pool their resources to construct a master file of 257 articles on the generally successful use of diethylstilbestrol. Ayerst Labs joins the group effort as a "dress rehearsal" for the subsequent approval of Premarin.
- In May, Frailey sends a letter to the consortium of drug-makers to report good news: "The time now seems propitious to suggest that you

re-file your new drug application for stilbestrol . . . the suggestion offered has official background."

- September: Permission to market diethylstilbestrol for treatment of
menopause symptoms is approved by the FDA.
- The father of metabolic bone research, Harvard's Dr. Fuller Albright,
publishes a paper in the *JAMA* called "Postmenopausal Osteoporosis:
Its Clinical Features." He proposes that diethylstilbestrol might be a
stimulus for bone formation. This is the first recommendation for long-
term therapy from a prominent scientist.

1942

- In May, the FDA grants permission to market Premarin for the treatment of menopause symptoms. As Premarin has already been prescribed for thousands of Canadian women, it is likely that some clinicians would have noted the endometrial disturbances that are visible in
10 percent of users by the end of the first year.

1943

- Dr. Robert Greenblatt, an early innovator in hormone delivery systems
such as estrogen and protestin pellets placed under the skin, publishes
two electrifying articles, one in the *JAMA* and the other in the *Journal
of Clinical Endocrinology*, on the benefits of testosterone pellets im-
planted in fifty-five female patients. In the endocrinology journal he
argues that testosterone use leads to a return of "coital pleasure" and
orgasm.

1947

- The FDA approves diethylstilbestrol use during pregnancy for the prevention of miscarriage. Behind this move was Dr. George Van Siclen
Smith, chief of the gynecology department at Harvard from 1942 until
1967, and his wife, the biochemist Olive Watkins Smith. (In 1976, five
years after the tragedy surfaced of rare cancers and reproductive organ
abnormalities in "DES babies," I would ask the Smiths whether they
were aware of the 1930s evidence of cancer and intersexuality pro-

duced in the babies of laboratory animals exposed to diethylstilbestrol. George Smith replies: "You can do all sorts of things to rats and mice by giving them overdoses.")

- Dr. Saul Gusberg, a young gynecologist and cancer researcher at the Sloane Hospital and Columbia University in New York, sounds an alarm. In an article published in the December 1947 issue of the *Journal of Obstetrics and Gynecology,* he reports that the relatively low cost and ease of administration of estrogen replacement therapy (ERT) has made its general use "promiscuous." He describes twenty-nine cases that he has followed that show that ERT overstimulates the endometrium, producing a "crowding of the glands into a lawless pattern." Twenty patients had a possibly premalignant condition called hyperplasia, while nine already had full-blown cancer.

- The first edition of *Physicians' Desk Reference* is published. It lists fifty-three different formulations sold by twenty-three companies for "menopausal disorders."

1948

- Dr. William Masters, later to coauthor the best-seller *Human Sexual Response*, publishes his first article, to be followed by a dozen more, on the effects of HRT on aging women. Masters, whom the science historian Elizabeth Siegel Watkins (2004) calls "one of the architects of the campaign to expand the use of sex hormones from short-term remedy . . . to long-term therapy," attributes the onset of osteoporosis, cardiovascular disease, and senility to the decline of estrogen production. Watkins maintains that "small clinical studies without proper controls or statistically significant sample sizes provided the rationale for the widespread practice in the following decades of prescribing long term hormone replacement for women."

1951

- The first edition of Madeline Gray's *The Changing Years* is published. Gray empowers women to have more say in whether to use estrogens, teaching her readers which symptoms they can help, the questions to ask doctors, and the risks of staying on hormones too long.

1966

- Dr. Robert Wilson, a general practitioner from Brooklyn, New York, publishes *Feminine Forever,* asserting that "while not all women are affected by menopause to this extreme degree, no woman can be sure of escaping the horror of this living decay." Ayerst Labs not only finances the book and provides a ghost writer, it systematically buys up retail copies to assure its appearance on best-seller lists.
- In November 1966, Wilson receives a sharp slap on the wrist from the FDA, which pronounces him unacceptable as an investigator of drugs for the treatment of menopause because he was "disseminating promotional material claiming that the drug had been shown to be effective for conditions for which it had never been proved to work." Wilson also claims that "all post menopausal women are castrates" and that hormones "prevent breast and genital cancer." He calls menopause a "deficiency disease like diabetes," insisting that estrogen will keep a woman "sexually attractive and potent," preserving the strength of her bones, the glow of her skin, and the gloss of her hair. The FDA pronouncement means drug companies cannot submit Wilson's research to the FDA when they seek to get product approval. The other drug companies that funded the Wilson Foundation—Searle and Upjohn—withdrew, but American Home Products stayed on.

1969

- Another doctor fueling women's fears is the psychiatrist David Reuben, who in 1969 will gain fame as the author of *Everything You Always Wanted to Know about Sex.* In that book, he claims that "as the estrogen is shut off a woman comes as close as she can to being a man."

Mid-1960s–mid-1970s

- Estrogen sales double and triple. Contributing factors include the advent of Medicare and the establishment of the Information Center for Mature Women, a "service for media" provided by Ayerst Labs and located in New York City. In a style that is breezy and readable but authoritative, the executive director, Sondra Gorney, provides magazines, newspapers, and radio and television programs with up-to-date information about menopause questions and controversies at no charge. Many use excerpts from her newsletters and position papers, but they rarely mention that the center is a front for industry.

1975

- The *New England Journal of Medicine* publishes a series of articles showing that the risk of endometrial cancer in estrogen users is five to fourteen times higher than in nonusers. Two weeks later, on December 16, Ayerst sends a "Dear Doctor" letter about Premarin to physicians across the United States. The letter is reassuring, making it appear that the *NEJM* articles reported "weak studies" and that the link between estrogen treatments and cancer has not really been established. In a rare move, Alexander Schmidt, the FDA commissioner, describes the letter as "misleading and irresponsible." The statement is headline news and a severe rebuff to the company. The FDA orders the labeling changed to state both the potentially lethal effect and that the risk is of a very high order.

1976

- In August, Dr. Robert Hoover and his associates publish a study in the *New England Journal of Medicine* that reports that fifteen years after commencing therapy, ERT patients have twice as many instances of breast cancer as women who have never used it.
- In his influential book, *Office Endocrinology,* Dr. Robert Greenblatt of the Medical College of Georgia urges his colleagues to include progestins in their prescriptions for women who have not had a hysterectomy. Including progestins counters the cellular build-up in the uterus from estrogen alone.

1976–1978

- According to an FDA mandate, by April 1978 all estrogen products and birth control pills are supposed to contain a comprehensive warning to users, to be dispensed by the pharmacist at the point of sale. The Pharmaceutical Manufacturers Association, supported by both the

American Medical Association and gynecologists' organizations, seeks injunctions to block distribution of the patient information leaflets. Ultimately, a U.S. District Court in Delaware rules in the FDA's favor, after the Center for Law and Social Policy in Washington, D.C., files a brief on behalf of several women's rights and consumer organizations. The leaflets make clear that estrogen has been proved effective only for hot flashes and vaginal dryness while carrying risks of cancer and blood clots. Estrogen use falls substantially, to six million women or fewer by the early 1980s, and most of these users were on it for the short term, no more than a few months to a year or two, to control symptoms.

1978

- A government task force to study diethylstilbstrol (known as the "DES task force") reports that women who took the drug to prevent miscarriage have elevated breast cancer rates, while some of their daughters develop a rare vaginal cancer that can be lethal; many daughters and some sons have fertility problems and other reproductive abnormalities that recall the "experimental inter-sexuality" described in animal research in the 1930s. The task force, on which this author sat, recommends government supported screening for all DES offspring and the removal of DES from animal feed.

1982

- A San Francisco radiologist named Harry Genant publishes what is now a classic paper in the *Annals of Internal Medicine.* Genant picks up on the 1941 metabolic "bone formation" report by Fuller Albright. Genant recruits twenty-seven volunteers, all of whom were still menstruating when their ovaries were removed. In a randomized, double-blind clinical trial he shows that patients who received an intermediate dose of Premarin retained their bone mass far better than those on lower doses or placebos. This rather technical article is entitled "Qualitative Computed Tomography of Vertebral Spongiosa: A Sensitive Method for Detecting Early Bone Loss after Oophorectomy." When the article is summarized and publicized in the mass media, the stories fail to mention, the central fact that not one of the twenty-seven volunteers was a woman going through a natural menopause at the usual age. Women, doctors, and anyone who missed the actual article were misled. Similar omissions, called selective disclosure, would occur frequently for the next two decades, contributing to the revival of estrogen as a must-have for healthy bones, heart, skin, sex life, and brain.

1984

• The National Institutes of Health convenes the Consensus Development Conference on Osteoporosis, chaired by Dr. William Peck, professor (later dean and eventually president) at the Washington University School of Medicine in St. Louis. The report from the conference urges that low-dosage bone density–measuring devices, less expensive than Dr. Genant's state-of-the-art research-quality equipment be developed and made readily available. A rush to get FDA approval follows in which questionable devices are "grandfathered" in, without testing, based on similarities to earlier equipment that was never reliable in the first place. Women are subjected to aggressive marketing and exposure to a decade of unreliable and sometimes dangerous bone-measurement devices.

1985

• The campaign to restore long-term use of hormones for the prevention of bone loss is meticulously put in place. Ayerst hires Burson-Marsteller, a public relations firm, to conduct a survey, that finds that 77 percent of women have never heard of osteoporosis. It introduces an "educational campaign" to position osteoporosis as an unrecognized killer disease. The campaign initiates a National Osteoporosis Week and places articles in numerous newspapers and magazines, including *Vogue, McCall's,* and *Readers' Digest.*

• From the Harvard School of Public Health, the ambitious Nurses' Health Study reports in the *New England Journal of Medicine* that among its 32,300 postmenopausal women, hormone users had half the rate of coronary disease compared with those who never took the hormones. However, in the same issue of *NEJM*, researchers at the Framingham Heart Study reveal that estrogen users have twice as many strokes and blood clots and a 50 percent *higher* coronary risk. The bad news from the Framingham study elicits far less discussion and media attention than the good news from the Nurses' Health Study. This becomes part of a larger trend where positive news about hormone drugs is touted in the front pages of newspapers, and negative findings are buried further in. Indeed, as "Science by Press Release" picks up momentum, the quality of a study seems to become less important than the commercial prospects of the message.

With the advent of direct-to-the-consumer advertising in the print media by the late '80s and early '90s, and on television in the late '90s, as Gloria Steinem puts it: "If you don't praise their products they won't buy your ads."

1985–86

- Following his successful NIH Consensus Conference, Dr. William Peck raises drug industry support and becomes founding president of the National Osteoporosis Foundation, a position he holds until 1990. Dr. Peck, a Washington University professor, is an authority on bone formation and a widely influential scientist, having written more than one hundred articles and edited many related journals and textbooks. He has chaired the endocrine and metabolism FDA Advisory Committee, NIH study sections, and has served as president of the American Society for Bone and Mineral Research. Soon after becoming the head of the National Osteoporosis Foundation, he becomes dean of his medical school and later chair of the national association of medical deans. His brilliant career permits him immeasurable influence on what osteoporosis research will be published or discarded, what drugs in the field will be approved for which indications, and what standards medical schools will require in their research contracts with industry. When this reporter asks him if it is true that some bone densitometry equipment is not reliable, he replies that one brand is "very good" but he cannot tell me which brand as he is a director of the company. Thanks to Dr. Peck's great knowledge and connections, the FDA adds "prevention of osteoporosis" to hot flashes and vaginal dryness as a confirmed reason for "keeping her on Premarin." Dr. Peck is an important example of the new relationship between university and NIH and drug companies. As former New England Journal of Medicine editor Marcia Angell explains in her 2004 book, The Truth About the Drug Companies, after the Bayh-Dole Act of 1980, universities were able to use tax dollars to create drugs and medical devices which they could then sell exclusively to drug companies.

1989

• A report from Sweden involving twenty-three thousand women indicates that the addition of progestin appears to increase the risk of breast cancer to at least twice the risk from estrogen alone. This could have and should have made doctors and women surmise that long-term users *who had not had hysterectomies* might be facing a difficult choice: don't add progestin and their odds of suffering cancer in the uterus increase by as much as twentyfold; add the progestin and they might double or triple their risk of breast cancer compared to taking estrogen alone. In 2000–2003, the National Cancer Institute, the Women's Health Initiative (WHI), and the "Million Women" breast cancer study in England would confirm that the combination of estrogen and progestin, while having a protective effect on the uterus, is more dangerous to the breast.

1990

• Premarin's manufacturer, then called Wyeth-Ayerst (now called Wyeth), loses its bid for FDA approval for claims that the drug prevents heart disease. Congress approves funding for the Women's Health Initiative (see Wassertheil-Smoller, this volume), after a few determined people, including Cindy Pearson of the National Women's Health Network, cry sexism, informing the FDA staff that no heart-protective drug, including aspirin, has been approved for males without evidence from serious clinical trials. At the same meeting where Pearson makes this argument to the FDA, Dr. Elizabeth Barrett-Connor points out that the evidence of estrogen's heart protective capabilities could be merely observational. Estrogen users are likely to be better educated, more affluent, nonsmokers, taller, thinner, and healthier to start with, which means they have fewer risk factors for heart disease.

1990s Throughout

• Pharmaceutical industry–supported groups flourish by masquerading as independent professional or grassroots consumer organizations.

These include not only the Osteoporosis Foundation but also the North American Menopause Society (NAMS) and American Council on Science and Health (ACSH). While these organizations may perform worthwhile services, and support some good research, they also promote and defend the corporations that sustain them. Sixty percent of the continuing medical education courses that U.S. physicians are required to take in order to maintain their licenses enjoy the sponsorship of the pharmaceutical companies, which strongly influence the content.

• Starting in the mid-1990s, Dr. Drummond Rennie of the *Journal of the American Medical Association* begins to speak out and write editorials against the hidden economic entanglements of drug companies with medical research and education. He is joined by editors at the *New England Journal of Medicine,* and today most high-quality science publications require that all contributors supply potential conflict-of-interest information. The media do not consistently convey this information to the public.

1996

• At an FDA workshop on patient information leaflets, AMA delegate Roy Schwarz admits that his organization opposes detailed labeling of hormone products because its members do not want patients to know which usages are "off-label." This term refers to drugs that are approved by the FDA but not for the specific condition at hand. Examples would be prescribing estrogen for heart protection or Alzheimer's prevention. These benefits were never proven.

• In England, the *Lancet* publishes three articles confirming that among healthy women aged forty-five to sixty-five, potentially fatal blood clots in the legs and lungs increase twofold to fourfold on estrogens and estrogen/progestin regimens. At this time Premarin is the best-selling prescription drug in the United States, yet only six U.S. newspapers in the Lexis Nexis database report the findings, and then, only in a sentence or two in "roundup" columns. Not until 2002, with the halting of the WHI Prempro trial, did most U.S. women hear about this risk.

2002

• The WHI halts the Prempro trials due to increases in breast cancer, heart disease, stroke and blood clots. There was a decline in colon cancer and fractures but not enough of a difference to compensate for the risks.

- In response to the halting of the Prempro trial, the eminent Canadian epidemiologist Dr. David Sackett writes an acclaimed editorial called "The Arrogance of Preventive Medicine." He notes, "Preventive medicine displays all three elements of arrogance. First, it is *aggressively* assertive, pursuing symptomless individuals and telling them what they must do to remain healthy. Second, preventive medicine is *presumptuous,* confident that the interventions it espouses will, on average, do more good than harm to those who accept and adhere to them. Finally, preventive medicine is *overbearing,* attacking those who question the value of its recommendations." In expressing his admiration for the volunteers who placed their bodies on the line, Dr. Sackett writes: "First place among the heroes is shared by each of the 16,608 women who agreed to collaborate in the estrogen-progestin portion of the Women's Health Initiative randomized trial."

- On September 18, 2002, only two months after the announced decision to stop the WHI Prempro trial, a group of five eminent academic specialists, led by Dr. Nathan Kase, former chair of obstetrics and gynecology at Yale, sends a letter to gynecologists urging them to switch their patients from oral hormone products to estradiol patches. Contrary to the tone of the letter, the actual studies, so far, comparing women who use the patches with those on pills, indicate that breast cancer and heart risks are not lower in patch users. The jury is still out regarding blood clots where the study results are mixed. In 2002, Dr. Kase is chairman of the *Women First* advisory board. *Women First* is a specialty pharmaceutical company based in San Diego that features Esclin, a brand of estrogen patch.

2003

- The FDA orders a "black box" warning on estrogen products. This label emphasizes that these products should only be used for specified symptoms, particularly hot flashes and night sweats, and used for the shortest possible duration.

2004

- A ten-year trial of 247 women with osteoporosis, published in the *New England Journal of Medicine,* establishes that those taking Fosamax maintained their bone mass about 10 percent better than those on the placebo. However, according to Dr. Abby Lippman, co-chairwoman of the Canadian Women's Health Network and a professor of epidemiology and biostatistics at McGill University, the study results failed to demonstrate either a reduction in fractures or better maintenance of height. Meanwhile, a striking change is taking place as many researchers and health officials are reconsidering old-fashioned, non-drug solutions, such as fracture and fall prevention, emphasizing appropriate exercise regimens and common sense solutions like making sure staircases have banisters, pot holes are repaired, and seniors avoid unsafe rugs and home furnishings.

- Also in March, NIH safety monitors end the WHI clinical trial of Premarin alone. Women and their doctors step up their search for alternatives.

This timeline reflects the many years of research that Barbara Seaman has done on this topic. The list of sources that follows is only partial. Further documentation can be found in Seaman's The Greatest Experiment Ever Performed on Women.

Partial Bibliography and Sources

1890s
Merck Manual Diagnosis and Therapy (New York: Merck, 1899).

1922
Edgar Allen sources at Yale University School of Medicine Library, particularly his last paper before his untimely death, "Estrogenic Hormones in the Genesis of Tumors and Cancers." *Endocrinology* (June 1942).

1930s
Gary L. Nelson, ed. 1983. *Pharmaceutical company histories*, Vol. 1: *Ayerst laboratories* (Bismarck, N.D.: Woodbine Publishing.)
Wolfgang Frobenius. 1990. *A triumph of scientific research: The development of ethinylestradiol and etjinyltestosterone, a story of challenge overcome* (England: Parthenon).
See also writings on Butenandt by the historian Robert Proctor, particularly (2000) *Adolph Butenandt, 1903–95.* Published in German. (Berlin: Presientenkcommission). See also Benno Muller-Hill, 1998. *Murderous science—Elimination by scientific selection of jews, gypsies and others in germany, 1933–45* (New York: Cold Spring Harbor Press). Originally published in German in 1984.

1938   E. C. Dodds. 1938. *Nature* (February 15).
Barbara Seaman. 2003. *The greatest experiment ever performed on women* (New York: Hyperion). See chapter 2, Fountain of youth or golden fleece? and chapter notes, pp. 31–43 and 291–95.
For initial reports on hot flashes see L. F. Hawkinson. 1938). The menopausal syndrome; One thousand consecutive patients with estrogen. *Journal of the American Medical Association* 11:390–93.
Weisbader and Kurzork. 1938. The menopause: A Consideration of the symptoms, etiology, and treatment by means of estrogens. *Endocrinology* 23: 32–38.

1940s
1939   Charles Geschickter. 1939. *Radiology.*
1940   J. R. R. Greene, V. W. Burrill, and A. C. Ivy. 1940. Experimental intersexuality. The effects of oestrogens on the antenatal development of the rat. *American Journal of Anatomy* 67: 305–45.
1940   Michael B. Shimkin and Hugh L Grady. 1940. *Proceedings of the Society for Experimental Biology and Medicine* 45: 246–48, Michael B. Shimkin and Hugh L Grady. 1940–41. *Journal of the National Cancer Institute* 1:119–27.
Copies of Chen and Harris report, and early DES studies available from Pat Cody at DES Action Archives. http//www.DESAction.org
1941   On the campaign to secure FDA approval, see Robert Myers (1983), *D.E.S: The bitter pill* (New York, Seaview/Putnam).
1941   F. Albright, P. Smith, and R. Richardson. 1941. Postmenopausal osteoporosis: Its clinical features. *Journal of the American Medical Association* 117:2473–76.
1941   R. B. Greenblatt. 1941. *Journal of the American Medical Association* 121:17.
R. B. Greenblatt. 1943. Hormone factors in libido. Editorial. *Journal of Clinical Endocrinology* (May): :305–6.
FDA records—U.S. government DES Task Force transcripts. 1978.
1947   S. B. Gusberg. 1947. Precursors of corpus carcinoma: Estrogen and andenomatojus hyperplasia. *American Journal of Obstetrics and Gynecology* 54:905–26.
1948   Elizabeth Siegel Watkins. 2004. The neutral gender and the problem of aging. Abstract. Annual Meeting of American Association for the History of Medicine.

1950s
1958   Madeline Gray. 1958. *The changing years* (New York: Doubleday).

1960s
Robert Wilson, 1966. *Feminine forever* (New York: M. Evans).
David Reuben. 1969. *Everything you always wanted to know about sex but were afraid to ask.* (New York: Rawson).
1966   On the popularity of estrogen products see Barbara Seaman (1972), *Free*

and female (New York: Coward, McCaan & Geoghegan); Barbara Seaman and Gideon Seaman. M.D. (1977), *Women and the crisis in sex hormones* (New York: Rawson, 1977); Barbara Seaman (2003), *The greatest experiment ever performed on women: Exploding the Estrogen Myth* (New York: Hyperion).

1970s
1976   Harry Zeil and William Finkle. 1976 "Association of Estrone with the Development of Endometrial Cancer," *American Journal of Obstetrics and Gynecology* 124: 735.
1976–78   Records at Center for Law and Social Policy, Washington, D.C., National Women's Health Network, Washington, D.C.; FDA records.
1978   Task Force Proceedings.

1980s
1982   Harry Genant. 1984. Quantitative computed tomography of vertebral spongiosa: A sensitive method for detecting early bone loss after oophorectomy. *Annals of Internal Medicine.*
1985–89   National Institutes of Health, Consensus Development Conference on Osteoporosis, chaired by Dr. William Peck. 2000. 100 leaders for the millennium,. *St. Louis Business Journal,* January 28; Seaman, *Greatest Experiment,* pp. 102–3n. 72.
1985–86   Marcia Angell. 2004. *The truth about the drug companies* (New York: Random House).

1990s
1990   Cynthia Pearson and Elizabeth Barrett-Connor. 1990. Transcripts of proceedings, Meeting of the Fertility and Maternal Health Drugs Advisory Committee/Center for Drug Evaluation and Research, U.S. Department of Health and Human Services, Food and Drug Administration, Gaithersburg, Md. Miller Reporting Co., June 15, 1990. See also Seaman, *Greatest Experiment;* Barbara Seaman. 1997. The media and the menopause industry. EXTRA! (March–April); Sharon Brownless. 2004. Doctors without borders: Why we can't trust medical journals anymore. *Washington Monthly* (April); Leora Tannenbaum. 1999. The bitter pill. In *For Women Only!: Your guide to health empowerment,* ed. Barbara Seaman and Gary Null. (New York: Seven Stories Press), 1379–84.

2000s
Jocalyn Clark. 2003. A hot flush for big pharma: How HRT studies have got drug firms rallying the troops. *British Medical Journal* (August 16): 400; Klim McPherson and Elina Hemminki. 2004. Synthesizing license data to assess drug safety./ *British Medical Journal* (February 28); C. David Naylor. 2004. The complex world of prescribing behavior. Editorial. *Journal of American Medical Association* 291

(January 7): 104–6; Kelly Hearn. 2004. "The drug profiteers." www.alternet.org. August 13;

Report of the NIH Blue Ribbon Panel. 2004. Conflicts of interest: A Working Group of the Advisory Committee to the director of the NIH, June 22. www.nih .gov/about/ethics_COI_panelreport.pdf.

14

Symptom Reporting at the End of Menstruation

Biological Variation and Cultural Difference

Margaret Lock

Introduction

It is well recognized that the end of female reproductive life is a complex transition that involves not only biological but also psychological and social changes. However, it is usually assumed that the biological changes that take place at this time, externally evident by the cessation of menstruation, are essentially universal and that differences reported by individual women can be fully accounted for by the numerous psychological, social, and cultural factors that shape subjective experience and are, in effect, layered over an invariant biological base (for notable exceptions to this argument see Kaufert, Gilbert, and Tate 1989; Mansfield and Jorgensen 1985; and Voda 1981).

Female reproductive senescence is, of course, universal but recent research strongly suggests that, in addition to psychosocial and cultural differences in women's experiences, considerable biological variation is implicated in the menopausal experience. These findings in turn mean that it is appropriate to think of biology and culture as in a continuous exchange in which *both* factors are subject to variation, rather than understanding the biology of menopause as a uniform foundation over which cultural difference is layered.

Researchers have shown that the incidence of symptoms taken as characteristic of menopause in North America and Europe, in particular hot flashes and night sweats, are not distributed equally among populations of peri- and postmenopausal women (see Leidy Sievert 2001 and Obermeyer et al. 1999 for reviews of this literature).[1] Nor are postmenopausal women equally at increased risk of heart disease, osteoporosis, or other late-onset chronic diseases. The considerable variation in the incidence of these conditions, both within any given society and cross-culturally, is well established. This variation suggests that the blanket recommendations made until recently for long-term use of hormone replacement therapy (HRT) by all postmenopausal women were ill advised. These recommendations have recently been withdrawn on the basis of the findings from two major clinical trials—the Women's Health Initiative (WHI) study and the Heart and Estrogen/progestin Replacement Study (HERS)—that made clear that the risks associated with long-term use of HRT are greater than its benefits. This finding reinforces the cross-cultural data reported here, which show that the health ramifications of lowered estrogen levels are by no means similar for women everywhere.

Menopause as Cultural Construct

The idea of menopause—that is, the ways in which health care professionals and individual women conceptualize and talk about changes associated with the end of menstruation—vary significantly, both historically and cross-culturally. Today, the usual understanding of the word *menopause* in the medical world and among women in Europe, North America, and (increasingly) elsewhere is that it means the end of menstruation. Menopause is in effect conflated with the end of menstrual cycling. However, such thinking is very recent, having spread rather gradually, commencing from about the end of the nineteenth century, first among gynecologists, and then eventually to the public at large. Only in the latter part of the twentieth century did this understanding become "common sense" knowledge. However, in many parts of the world, where social transitions associated with aging are regarded as of prime importance and individual biological changes often attract less attention, if the meaning of *menopause* is confined rather closely to the end of menstruation, the word is being used in ways that do not "fit" well with local accounts about women as they age (see, for example, Bledsoe 2002 and Lock 1993).

In North America and Europe, in part because of the conflation of the concept of menopause with the end of menstrual cycling, the dominant understanding of menopause since the mid-1970s has been one of a diseaselike condition. Because the focus of attention is, above all, on declining estrogen levels, becoming postmenopausal has been likened to having a deficiency disease, similar to diabetes, for example, in which insulin is deficient. Implicit in this approach, glossed as the "narrow hypothesis of estrogen decline" (Obermeyer 2000,184), and closely associated with the experience of vasomotor symptoms, is the mistaken idea that menopause is a recent phenomenon in human history. Articles in many publications have suggested that because, until the turn of the twentieth century, mean life expectancy for North American and northern European women was less than fifty years of age, and even lower elsewhere in the world, virtually no women lived much past middle age until the recent past (see, for example, Sheehy 1992, Gosden 1985). Postmenopausal life, it is often claimed, is a cultural artifact—the result of better health care and medical services that have inadvertently brought about this deficiency state.

Such claims are very misleading. Until the first part of the twentieth century, infant mortality rates were high, and of those women who survived to reproductive age, many died in childbirth. If a woman lived through her own infancy and then her reproductive years, survival to old age was in fact common. In order to take this effect into account when calculating age distribution in human populations, it is essential to estimate remaining life expectancy once individuals have reached forty-five or fifty years of age. A closer look at these figures makes clear that living to old age is certainly not a recent phenomenon, although more people survive to old age today than was formerly the case.

The literature on menopause often makes a second point in connection with the presumed distinctiveness of postmenopausal women, namely, that virtually no mammals in the wild live past reproductive age. We do not yet fully understand the evolutionary advantage of a postmenopausal life in human females, but we do know that to live past reproductive age in women, although unique among mammals, is a very old human characteristic of many hundreds of thousands of years' duration. Biological anthropologists agree that menopause is genetically programmed, and one theory for its existence is the so-called grandmother hypothesis. This theory posits that postreproductive life in human females has been biologically selected for because investment by older women in the offspring of their children, that is, in their grandchildren, benefits the group as a whole. The assumption of this theory is that under these circumstances fewer infants die than if women were to continue to produce their own offspring throughout their entire life span without any help from others to provide the intense care and attention that human infants require. This hypothesis has yet to be proved, but even if incorrect or only partially correct, it simply cannot be argued that the postmenopausal condition is in effect "unnatural," a diseaselike condition that we have created for ourselves in modern life.

Factors Contributing to Variation in the Menopausal Experience

Using a life-span approach, Leidy (1994:240) has argued that although variation in age at the last menstruation is confined to a narrow spectrum, this variation is nevertheless of significance and is influenced by family history. She notes: "Genetically, parents pass to their daughters the parameters for number of oocytes and/or rate of atresia [degeneration of ovarian follicles]. Behaviorally, a mother's activity while pregnant affects the ovarian store her daughter possesses at birth. From birth until menopause the environment and behavior of the individual affects her own ovarian stores." Diet, age at menarche, reproductive history, use of the birth-control pill and of medication, and smoking history—to name the most obvious variables—are all implicated.

When it comes to differences in symptom reporting during the perimenopausal years, in addition to individual biological variation and dietary differences, other factors must be taken into consideration. Included are language usage and whether the end of menstruation is culturally marked, as well as the symbolic meanings, cultural stereotypes, and preformed expectations about this stage of the life cycle, and individual self-esteem and social role transitions, all of which function to modify the subjective experience of the menopausal transition.

I want to emphasize that, no matter where the research has been carried out, it always shows that the majority of women pass through the menopausal transition with relatively little or no discomfort. Survey research in the United States and Canada with large samples of women aged forty-five to fifty-five who are representative of the general population add substantial support to this finding (McKinlay, Brambilla, and Posner 1992; Kaufert, Gilbert, and Hassard 1988). Obermeyer (2000) reinforces this point in a review article, but she nevertheless notes that uncomfortable symptoms are readily associated with menopause in virtually all societies that have been investigated. These negative experiences, although they are in the minority, become the central element of a dominant negative stereotype even where medicalization of menopause is not the norm. This finding is not easily explained and demands more attention. But studies also have shown that people everywhere attach both negative and positive meanings to the end of menstruation: Qualitative research demonstrates the inevitable ambivalence associated with the end of reproductive life, the implications of which vary enormously depending upon local attitudes toward aging in general, older women in particular, and their place in society.

Cross-Cultural Research on Menopause

Cross-cultural research in connection with menopausal symptomatology has shown consistently that the incidence of symptoms taken as characteristic of menopause in North America and Europe, in particular hot flashes and night sweats, are not reported to the same extent among other populations of women as they go through menopause. Beyene (1986) found no reporting of either hot flashes or of cold sweats in rural Mayans living in the Yucatan, Mexico, where women have numerous pregnancies and extended cycles of amenorrhea (i.e. no menstrual cycles) associated with prolonged lactation and malnutrition. This was in contrast to the Greek peasant women whom she also studied, whose reported symptoms form a pattern similar to that in northern Europe. Beyene concluded that menopause should be understood as a biocultural event and emphasized that variables that must be taken into consideration include diet and reproductive history (Beyene 1986).

Research carried out in India (Flint 1975), Indonesia (Flint and Samil 1990; Boulet et al. 1994), among Africans living in Israel (Walfish, Antonovsky, and Mao 1994), in Taiwan (Yeh 1989; Boulet et al. 1994), Hong Kong (Haines, Chung, and Leung 1994; Boulet et al. 1994), Japan (Lock 1993), Singapore (McCarthy 1994; Boulet et al. 1994), China (Shea 1998;

Figure 14.1 A Samburu woman aged about forty-five, mother of
ten children (northern Kenya). Photo by Richard Lock.

Tang 1994), Korea (Boulet et al. 1994), Thailand (Chompootweep et al.
1993; Chirawatkul and Manderson 1994), and Malaysia (Ismael 1994) all
found low reporting of hot flashes and night sweats, although they also
found considerable variation in incidence of symptoms reported. Some
researchers have suggested that the women in these studies in effect bias
the research findings because they do not pay careful attention to their
bodies and, moreover, because individual responses to menopause are
culturally dependent, inattention to hot flashes is essentially learned
behavior.

For example, Boulet and colleagues (1994), on the basis of research in

Figure 14.2   A Samburu midwife, aged about fifty-five (northern Kenya). Photo by Richard Lock.

seven southeast Asian countries, showed that headaches, dizziness, anxiety, irritability, and other nonspecific symptoms were commonly associated with the menopausal transition by the women recruited into their project. In Japan, the symptom reported more than any other is shoulder stiffness (Lock 1993). Boulet and her associates argued that such symptom reporting should be understood as "a form of communication" on the part of women and speculated that vasomotor distress may be "translated" by women into culturally meaningful nonspecific symptoms that are associated with feelings of psychological distress. The assumption in making such interpretations is that when subjective reporting does not coincide with findings that the researchers anticipated, the women subjects are, in effect, being duped by their language and culture.

Much better standardization of research methodologies is necessary before anyone can reach firm conclusions on the basis of the findings of studies such as those cited earlier. Sample sizes are often small, and some are drawn from clinical populations, introducing considerable bias. Sensitivity to the complexity of linguistic expressions and translation of bodily terms across languages are often absent. Women who have undergone surgical menopause are not always placed in a separate category for the purposes of analysis. Further bias is introduced if, as is often the case, symptom reporting by women is not correlated with their actual menopausal status (something that is particularly difficult to establish where giving birth to many children is the norm). Very often women are asked to recall what may have happened five or even ten years previously. Despite these serious shortcomings, as a whole this research strongly suggests that considerable variation exists in the reporting of vasomotor symptoms across cultures and that it cannot be explained away as culturally conditioned inattention.

In a recent study with a sample of more than two hundred Mayan women whose average age at menopause was 44.3, endocrine changes at menopause were found to be very similar to those of American women of the same age, yet the Mayans, as noted earlier, report no hot flashes (except very occasionally after migration to an urban environment; Beyene and Martin 2001). It is known that plasma, urinary, and vaginal levels of estrogens do not correlate neatly with subjective reporting of hot flashes (Martin et al. 1993), nor do measured rates of sweating, peripheral vasodilation, and deregulation of core body temperature (Freedman 2001). Clearly, considerable mediation takes place between measurable physiological changes, subjective experience, and the reporting of symptoms, some of which may be accounted for by as yet poorly understood biological pathways (Kronenberg 1990; Ginsburg and Hardiman 1994). It is reasonable to speculate, for example, that with urban migration and education, women might experience hot flashes more frequently, perhaps as a result of dietary changes.

A Comparative Study

Analysis of comparable data sets comprised of 7,802 Massachusetts women, 1,307 Manitoban women, and 1,225 Japanese women aged forty-five to fifty-five, inclusively, reveal some remarkable differences in symptom reporting at menopause (Avis and McKinlay 1991). In the three research sites, samples were selected from a general and not a clinical population of women, and those women who had undergone gynecological surgery were placed in a separate category for analysis (table 14.1).

The Japanese word *kōnenki,* usually translated into English as menopause, does not convey the same meaning as *menopause,* which is usually understood as the end of menstruation. By contrast *kōnenki* is similar to the concept of the climacteric in that is it is thought of as a long, gradual process to which the end of menstruation is just one contributing factor. Japanese physicians deliberately created the term *kōnenki* at the beginning of the twentieth century as a result of close contact with German colleagues. Most Japanese respondents in my study placed the timing of *kōnenki* at age forty-five or even earlier, lasting until nearly sixty. One-quarter of the questionnaire respondents who were postmenopausal and had ceased menstruation for more than one year reported that they had no sign of *kōnenki.* It is notable that no word exists that refers uniquely to the hot flash in Japanese, even though this is a language that makes very fine discriminations in bodily states. Similarly, no local terms for the hot flash were found in China, Indonesia, the Yucatan, or in other locations.

In my study of North American and Japanese women (Lock 1993), I

Table 14.1 Percentage Distribution of Menopause Status in Each Study

Menopausal status	Study		
	Japan	Canada	USA
Natural menopause	35.7	35.5	43.9
Perimenopause	31.5	26.2	38.2
Premenopause	32.8	38.3	17.9
Surgical menopause	9.9	20.5	28.6
Total (100 percent)	1225	1307	7802

Note:
Totals exclude subjects with surgical menopause.

Figure 14.3   At work harvesting rice in northern Honshu, Japan. Photo by Richard Lock.

Figure 14.4 Taking mother-in-law for a walk, Kyoto, Japan. Photo by Richard Lock.

asked them to recall symptoms that they had experienced during the previous two weeks (research shows that longer periods of recall are very inaccurate). Japanese reporting of hot flashes was low, 13.5 percent for women whose menstruation had become irregular, and 15.2 percent for postmenopausal women. Although low in incidence, hot flashes are associated with the end of menstruation. For U.S. and Canadian women, approximately 39 percent and 42 percent of peri- and postmenopausal women, respectively, had experienced hot flashes in the previous two

Table 14.2 Reports of Vasomotor Symptoms by Menopausal Status

Menopausal status	Japan[°]	Manitoba[°°]	Massachusetts[°°]
Hot Flashes			
Premenopause	6.4	13.8	17.9
Perimenopause	13.5	39.7	38.1
Postmenopause	15.2	41.5	43.9
Total (100 percent)	1,104	1,039	5,505
	$x^2=15.77$	$x^2=84.17$	$x^2=269.510$
Night Sweats			
Premenopause	4.1	10.6	5.5
Perimenopause	4.0	27.6	11.7
Postmenopause	3.0	22.2	11.3
Total (100 percent)	1,104	1,039	5,484
	$x^2=0.772$	$x^2=33.71$	$x^2=31.335$

Note:
Massachusetts *N* differs because of missing data. Cases of surgical menopause have been removed.
[°]p = 0.00 and 0.68
[°°]p = 0.00

weeks. Reporting of night sweats was extremely low in Japan and not associated with menopausal status (table 14.2). Only 19 percent of Japanese women in this study had experienced a hot flash at some time in the past, and reporting of both frequency and intensity was much lower than among U.S. and Canadian respondents, nearly 60 percent of whom had experienced a hot flash at some time. Reporting of sleep disturbance by Japanese women, at 11.5 percent, was low, corroborating their reports about lack of severity of hot flashes. Follow-up interviews suggested that, if anything, the women in the Japanese study overreported such symptoms in their eagerness to cooperate fully with the researcher.

The majority of Japanese women, when asked in interviews to describe their experience of *kōnenki* responded along the following lines:

I've had no problems at all, no headaches or anything like that. . . . I've heard from other women that their heads felt so heavy that they couldn't get up.

The most common problems I've heard about are stiff shoulders, headaches, and aching joints.

I get tired easily, that's *kōnenki* for sure, and I get stiff shoulders.

A small number of women, twelve of the 105 interviewed, made statements that sound much more familiar to North Americans and Europeans:

The most noticeable thing was that I would suddenly feel hot; it happened every day, three times or so. I didn't go to the doctor or take any medication, I wasn't embarrassed, and I didn't feel strange. I just thought it was my age.

The questionnaire findings show that shoulder stiffness was the most common symptom reported in Japan, followed by headaches, and Japanese gynecologists note that women complain most often of these symptoms (more research supports these findings; see Kasuga et al. in press). During my research in the mid-1980s some physicians did not list hot flashes at all when I asked them to describe the symptoms of *kōnenki* (Lock 1993). Changes have occurred since that time: The majority of Japanese gynecologists are now interested in caring for women as they go through menopause, which was formerly not the case; articles about menopause appear frequently in Japanese women's magazines, where hot flashes are often described as the "typical" symptom of menopause (although clinical encounters indicate otherwise). These articles sometimes use a new word to describe this symptom—*hotto furashu*, taken directly from the English.

In responding to a questionnaire, 90 percent of Japanese gynecologists reported that they are willing to prescribe HRT. In addition, 70 percent of these same gynecologists also use herbal medicine to counter symptoms of *kōnenki* (Japanese Ministry 1996). However, it has been shown that only approximately 1.5 percent of women aged forty-five to fifty-four are actually prescribed HRT (Noji 2000). This suggests that, even before the stopping of the HRT trials in the United States in 2002, Japanese gynecologists either were not actively encouraging the use of HRT or, more likely, many Japanese patients continue to state a preference for taking herbal medicine over HRT, even though the Japanese are widely concerned about diseases associated with postmenopausal life. Another confounding factor is that the majority of Japanese women do not see a gynecologist at this stage of their life but instead choose to see an internist or general practitioner if and when they are actually experiencing discomfort.

It is of note that among those diseases associated with lowered estrogen levels, Japanese women and their physicians are particularly concerned about osteoporosis, stroke, and Alzheimer's disease, rather than heart disease. This reflects the relatively low rate of heart disease in Japan. People are particularly fearful about Alzheimer's disease, but concerns about the side-effects of long-term use of HRT apparently have outweighed a desire to use it as protection against Alzheimer's. This situation exists even though the Japanese media have done little reporting in Japan about the stopping of the clinical trials for long-term use of HRT in the United States or about the increasing agreement that HRT does not provide protection against Alzheimer's. The Japanese Menopause Society has recently started its own longitudinal study with Japanese research subjects. Japanese clinicians today comment frequently that results of research carried out with American subjects do not necessarily apply to Japanese women.

It is well known that Japanese women currently enjoy the longest life expectancy in the world—a mean of nearly eighty-two years. The incidence of breast cancer, and of heart disease in men and women, is about one-third of that in North America. The incidence of osteoporosis for both Chinese and Japanese women is less than half that of Caucasian women in North America, even though Asian women usually have a lower bone density. My research in Japan also showed that only 28 percent of Japanese respondents suffer from a chronic health problem (diabetes, allergies, asthma, arthritis, high blood pressure), compared with 45 percent of Manitoban women and 53 percent of Massachusetts women. Taken together, these figures suggest that middle-aged Japanese women enjoy better health than their counterparts in North America.

Women who were about fifty when this study was done were born at the beginning of World War II, and many experienced nutritional deprivation as very small children. However, virtually none of them has smoked; they consumed only small amounts of alcohol and coffee; and their diet is low in fat and rich in soy beans and vegetables. Soy beans, as is now well known, are a source of phytoestrogens and may well contribute to the less frequent reporting of hot flashes by Japanese women. So too may the herbal teas that many women drink, as some of these beverages also are rich in phytoestrogens. This cohort of women has, as part of their daily lives, always done considerable exercise and weight bearing. Obviously, this picture will change, as succeeding generations of Japanese women become middle aged in their turn.

Some additional findings of interest come from the recent work of

Jean Shea with four hundred Chinese women. She used the same research methods used in the Japanese and North American studies reported earlier. Vasomotor symptom reporting among the Chinese women is low and resembles that of the Japanese sample. However, the overall number of symptoms reported by the Chinese women is considerably higher than in the Massachusetts, Manitoba, and Japanese samples, leading Shea to conclude that sweeping generalizations about east Asia should be avoided (Shea 1998). Clearly, if the long-term health of aging women is of prime interest, factors such as relative deprivation, occupational stress, and political turmoil, such as Chinese women have suffered, must be factored into the equation.

Acknowledging Complexity

In attempting to account for differences in symptom reporting at menopause, it is helpful to draw on the concept of "local biologies" (Lock 1993). This concept does not refer to the idea that the categories of the biological sciences are historically and culturally constructed (although this is indeed the case) or to measurable biological difference across human populations, although such findings do contribute to the argument that I am making here. Rather, *local biologies* refers to the way in which the embodied experience of physical sensations, including those of well-being, illness, disease, and so on, are informed by the physical body, which is itself contingent upon evolutionary, historical, cultural, and individual variables.

Individual genomes vary as a result of changes that have taken place throughout both evolutionary time and the *longue durée* of historical change. These transformations are the result of interactions of human genes with the environments in which people have lived; their social arrangements, including marriage patterns and other arrangements that affect reproduction; and of the effects of specific behavioral patterns. The bodies of individuals represent a microcosm of these different effects.

But individual embodiment is also constituted by the way in which self and others represent the body, drawing upon local categories of knowledge and experience. If subjective bodily experience is to be made social, then history, politics, language, and local knowledge, including scientific knowledge to the extent that it is available, must inevitably be implicated. This means in practice that knowledge about biology is informed by social worlds, and the social world is in turn informed by the reality of physical experience. In other words, the effects of biology and the way in which it is understood and represented are virtually inextricable from one another, and the primary site where this engagement takes place is the individual body. One result is that the subjective experience of the end of menstruation for the majority of women in Japan, and in many other parts of the world, does not correspond at all well to what is assumed to be a universal event. Thus, although changes in ovarian and endocrine functioning are universally implicated in the midlife transition of women, beyond this biological fact, enormous complexity and variation come into play. This ranges from the possibility that endocrine functioning varies across populations to the possibility that the incidence of osteoporosis varies (Stini 1995) to the certainty that language used to describe female aging varies across cultures.

We would do well to acknowledge such variation rather than push it all to one side, regarding it as so much cultural noise that impedes the progress of science. Above all, to separate out the postulated effects of lowered estrogen levels from the aging process in general is to inappropriately transform an as yet poorly understood, complex biosocial process into one of a simple cause-and-effect relationship. This is not to suggest that certain women do not benefit greatly from appropriate medication but rather to highlight that we still have much to learn. To ignore the interdependence of biology, history, and culture is short sighted in the extreme.

Note

1. Women who are premenopausal have regular menstrual cycles. Perimenopausal women either have irregular menstrual cycles or have menstruated in the past twelve months but not in the previous three months. Women who have not had a menstrual cycle for more than a year are classified as postmenopausal. These are standardized ways of assessing menstrual status.

References

Avis, N.E., and S. McKinlay. 1991. A Longitudinal analysis of women's attitudes towards the menopause: Results from the Massachusetts Women's Health Study. *Maturitas* 13:65–79.

Beyene, Y. 1986. Cultural significance and physiological manifestations of menopause: A biocultural analysis. *Culture, Medicine, and Psychiatry* 10:47–71.

Beyene, Y., and M. C. Martin. 2001. Menopausal experiences and bone density of mayan women in Yucatan, Mexico. *American Journal of Human Biology* 13:505–11.

Bledsoe, C. 2002. *Contingent lives: fertility, time, and aging in West Africa.* Chicago: University of Chicago Press.

Boulet, M. J. et al. 1994. Climacteric and menopause in seven south-east Asian countries. *Maturitas* 19:157–76.

Chirawatkul, S., and L. Manderson. 1994. Perceptions of menopause in Northeast Thailand: Contested meaning and practice. *Social Science & Medicine* 39:1545–54.

Chompootweep, M. K. et al. 1993. The menopausal age and climacteric complaints in Thai women in Bangkok. *Maturitas* 17:63–71.

Flint, M. 1975. The menopause: Reward or punishment? *Psychosomatics* 16:161–63.

Flint, M., and R. Suprapti Samil. 1990. Cultural and subcultural meanings of the menopause, pp. 134–48. In *Multidisciplinary Perspectives on Menopause,* edited by M. Flint, Fredi Kronenberg, and Wulf Utian. Annual review of the New York Academy of Sciences, Vol. 592. New York: New York Academy of Sciences.

Freedman, R. R. 2001. Physiology of hot flashes. *American Journal of Human Biology* 13:453–64.

Ginsburg, J., and P. Hardiman. 1994. The menopausal hot flush: Facts and fancies, pp. 123–35. In *The Modern Management of the Menopause,* edited by G. Berg, and M. Hammer. New York: Parthenon.

Gosden, Roger R. 1985. *The biology of menopause: The causes and consequences of ovarian aging.* London: Academic Press.

Haines, C. J., T. K. H. Chung, and D. H. Y. Leung. 1994. A prospective study of the frequency of acute menopausal symptoms in Hong Kong Chinese women. *Maturitas* 18:175–81.

Ismael, N. N. 1994. A study of the menopause in Malaysia. *Maturitas* 19:205–9.

Japanese Ministry of Health and Welfare. 1996. Kōseishō survey of attitudes toward prescription of medication for menopausal patients. Tokyo: Ministry of Health (Kōseishō).

Kasuga, Michiko et al. In press. Relation between climacteric symptoms and ovarian hypofunction in middle-aged and elderly Japanese women presenting at a menopausal clinic. *Menopause.*

Kaufert, Patricia., Penny Gilbert, and T. Hassard. 1988. Researching the symptoms of menopause: An exercise in methodology. *Maturitas* 10:117–31.

Kaufert, Patricia, Penny Gilbert, and Robert Tate. 1989. Defining menopausal status: The impact of longitudinal data. *Maturitas* 9:217–26.

Kronenberg, Fredi. 1990. Hot flashes: Epidemiology and physiology. *Annuals of the New York Academy of Sciences* 592:52–86.

Leidy, L. 1994. Biological aspects of menopause across the lifespan. *Annual Review of Anthropology* 23:231–53.

Leidy Sievert, Lynnette. 2001. Menopause as a measure of population health: An overview. *American Journal of Human Biology* 13:429–33.

Lock, M. 1993. *Encounters with aging: Mythologies of menopause in Japan and North America,* Berkeley: University of California Press.

Mansfield, Phyllis Kernoff, and Cheryl M. Jorgensen. 1985. Menstrual pattern change in middle-aged women, pp. 213–25. In *Menstrual health in women's lives,* edited by Alice J. Dan and Linda L. Lewis. Champagne: University of Illinois Press.

Martin, M. et al. 1993. Menopause without symptoms: The endocrinology of menopause among rural Mayan Indians. *American Journal of Obstetrics and Gynecology* 168:1839–45.

McCarthy, T. 1994. The prevalence of symptoms in menopausal women in the Far East: Singapore segment. *Maturitas* 19:199–204.

McKinlay, S. M., D. Brambilla, and J. Posner. 1992. The normal menopausal transition. *Human Biology* 4:37–46.

Noji, Ariko. 2000. Quality of life for changing health: Perimenopause among Japanese in Japan and the USA, pp. 127–32. In *The Menopause at the Millennium,* edited by Aso Takeshi, Takumi Yahaihara and Seiichirou Fujimoto. Carnforth, Lancashire, U.K.: Parthenon.

Obermeyer, C. M. 2000. Menopause across cultures: A review of the evidence. *Menopause: The Journal of the North American Menopause Society* 7:184–92.

Obermeyer, Carla Makhlour, Françoise Ghorayeb, and Robert Reynolds. 1999. Symptom reporting at menopause in Beirut, Lebanon. *Maturitas* 33:249–58.

Shea, J. 1998. Revolutionary women at middle age: An ethnographic survey of menopause and midlife aging in Beijing, China. Ph.D. diss., Harvard University, Cambridge, Mass.

Sheehy, Gail. 1992. *The silent passage: Menopause.* New York: Random House.

Stini, William A. 1995. Osteoporosis in biocultural perspective. *Annual Review of Anthropology* 24:397–421.

Tang, G. W. K. 1994. The climacteric of Chinese factory workers. *Maturitas* 19:177–82.

Voda, Ann M. 1981. Climacteric hot flash. *Maturitas* 3:73–90.

Walfish, S., A. Antonovsky, and B. Mao. 1994. Relationship between biological changes and symptoms and health behavior during the climacteric. *Maturitas* 6:9–17.

Yeh, A. 1989. The experience of menopause among Taiwanese women. Honors thesis, Department of East Asian Languages and Civilizations, Harvard University, Cambridge, Mass.

15

Evidence-based Medicine and Clinical Practice

David L. DeMets

Introduction

The practice of medicine has evolved over many centuries, but the rate of that evolution has become much faster in recent decades. Clinical practice is increasingly influenced by the results of laboratory and clinical studies aimed at evaluating disease risk factors and methods of either preventing or treating disease. However, the results of these scientific studies do not always lead in a straight simple path to the best or optimum intervention and medical practice. First, the complexities of the human system make the task challenging. Second, not all results from the various studies give the same answer or clues so that drawing general conclusions is not simple and in some cases not even feasible. Finally, not all data or studies are equally valid and must be interpreted accordingly. Hormone replacement therapy for postmenopausal women is an illustration of these difficulties.

Types of Studies

Researchers obtain information from a variety of sources. Figure 15.1 presents an oversimplification that shows clinical observation, laboratory studies, and clinical experiments. These three sources of evidence are interrelated; research in each of these are typically required to develop successful new treatments.

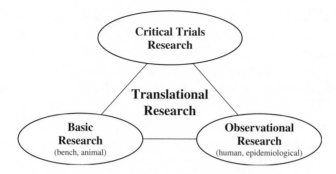

Figure 15.1   Three interrelated sources of evidence, research in each of which is typically required to develop successful new treatments.

Observational studies generally collect data by observing an individual person (a case study) or a group of people (case studies). The individuals may be healthy normal people, individuals at higher risk for a disease, or individuals recently diagnosed with a disease. The individual observation or single case study may identify a new medical problem. For example, in the mid-1990s astute observation by a clinical team identified a series of individual patients who have a heart valve abnormality with a frequency greater than expected from the general population. The clinicians noted that these patients were taking a weight-reducing combination, fen-phen (Connolly et al. 1997). This clinical observation led to further investigation, which confirmed the suspicion and ultimately caused the removal of the drug from the market. Such discoveries are rare, but they are extremely valuable when they occur. More commonly, data are collected systematically on large groups or cohorts of patients that may lead to new discoveries. Observational studies can be classified into three different groups: retrospective, cross-sectional, and prospective studies.

Retrospective studies typically collect data on patients who have been diagnosed with a disease in order to identify risk factors for the disease. In order to evaluate the potential risk factor, the same type of data is collected on individuals who do not have the disease. For example, patients diagnosed with lung cancer were evaluated as to their past behavior. Data were also collected on individuals who did not have lung cancer. Among the factors considered was smoking behavior. The data showed that patients with lung cancer were more likely to have had a smoking history than those who did not have lung cancer. These results were replicated in several studies, and researchers concluded that smoking indeed was a major risk factor for lung cancer (Wynder and Graham 1950; Doll and Hill 1954; Doll et al. 1994; U.S. Public Health Service 1964; Peto 1986). Such retrospective, or case control, studies have proved valuable in identifying risk factors for diseases with low incidence. Although such research designs are sometimes useful, they are also subject to biases that can mislead researchers about potential risk factors. For example, individuals may be selective in what they report about their past behavior or exposure to potential risk factors; this is referred to as recall bias. Choice of the control group population can also strongly influence the results. Thus retrospective designs can be useful in identifying potential risk factors but rarely are definitive.

Another type of design calls for collection of data on a group of individuals at a single point in time, a cross section of living individuals. By examining the types of disease present in this cohort and its members' medical history, physical characteristics, and other attributes, associations can sometimes be identified that lead to identification of new potential risk factors. More commonly, prospective studies are conducted in search of disease risk factors. Prospective studies characterize a cohort of subjects and then follow these people forward in time to collect data on disease incidence. One such well-known study was the Framingham Heart Study (FHS), which identified, for example, high serum lipid values and high blood pressure as risk factors for heart disease (Kagan et al. 1959; Kannel et al. 1962). However, the association of such parameters as serum lipids or blood pressure with heart disease does not mean that they cause heart disease, unless this is confirmed by further research. Many attributes can be associated with a disease but are not causal. For example, red meat consumption and breast cancer and colon cancer, or vasectomy and prostate cancer (Taubes 1995) are associations that have been reported but have not been proven to be causally related.

If we identify a risk factor such as serum cholesterol for coronary heart disease, we need to next find a method to modify that risk factor. Thus laboratories might experiment with drugs to find those that will lower serum cholesterol, first in animals and then in humans with high cholesterol levels. Once a drug is found that effectively lowers serum cholesterol, the researchers must conduct a clinical study to determine whether lowering cholesterol actually results in a reduction in the incidence of coronary disease. For example, several drugs that lowered serum cholesterol were tested in the Coronary Drug Project but not all reduced the incidence of coronary disease morbidity or mortality (Coronary Drug Project 1975). After two decades of research, no cholesterol-lowering strategy demonstrated a substantial reduction in coronary heart disease mortality or serious morbidity. Ultimately, however, a class of drugs called statins, which effectively lowered serum cholesterol, was proved to also effectively lower the incidence of coronary heart disease (Scandinavian Simvastatin Survival Study Group 1994; Shepherd et al. 1995).

Often data from clinical trials can be used to identify new risk factors or groups at especially high risk. This discovery can lead to new studies or modifications of existing treatments. Results of trials can also suggest improvements in the drug being tested. For example, a drug might effectively modify the risk factor and reduce the disease incidence but have too many side-effects. Laboratory research might attempt to modify the molecular structure of the drug to maintain the effectiveness while reducing the side-effects. Alternatively, delivering the drug in a sustained release over a twenty-four hour period instead of all at once may reduce the side-effects while delivering the same effective dose.

This research paradigm forms the basis for evidence-based medicine. Not every step is taken in sequence, and some steps are skipped altogether. Evidence accumulates piecemeal, and how such evidence affects clinical practice cannot always be predicted. Many effective medicines produce a small but important improvement, such as a 10 to 20 percent reduction in the rate of death or recurrence of the disease. In order to discover and confirm that these small but clinically important improvements are real, the randomized clinical trial has become the standard by which we judge new interventions (Friedman, Furberg, and DeMets 1998). As I described earlier, this approach has been used successfully quite often, for example, for maintaining tight glucose control for diabetics (Diabetes Control and Complications 1993), lowering blood pressure for mild hypertension (Hypertension Detection and Follow-up Program 1979), using statins for lowering serum cholesterol and coronary heart disease (Scandinavian Simvastatin Survival Study Group 1994; Shepherd et al. 1995), and prescribing tamoxifen for reducing the recurrence of breast cancer in early-stage patients (Early Breast Cancer Trialists 1998) or for reducing the risk of breast cancer (Fisher et al. 1998).

Even with the benefit of a randomized control clinical trial, the interpretation of results is not always straightforward. For example, the tamoxifen primary prevention trial, comparing tamoxifen to a placebo in a population of women without breast cancer but at higher risk for breast cancer based on factors such as age and family history, was terminated early with a favorable and early effect in reducing breast cancer in the tamoxifen-treated patients (Fisher et al 1998). A statistically insignificant reduction in mortality also was observed among those taking tamoxifen. Interpretation of these results has been somewhat challenging. As the authors acknowledge, it is not clear whether tamoxifen prevented the initiation of breast cancer and the elimination of tumors or reduced the occurrence of previously undetected breast cancer. Furthermore, the results could not be confirmed in two other trials (Powles et al. 1998; Veronesi et al. 1998). Finally, because the trial was terminated early and thus the small number of events, the risk of fractures of the hip, spine, and wrist, and of cardiovascular disease morbidity and mortality among those taking tamoxifen remains unknown. In fact, treatments such as tamoxifen affect many systems, and the final evaluation must balance the risks and benefits across all these dimensions.

However, there are also examples where this paradigm was not used and inappropriate treatments came into practice, such as, for example, the use of drugs that suppress cardiac arrhythmia in patients with irregular heartbeats (Cardiac Arrhythmia Suppression Trial Investigators 1989), the use of intermittent pressure breathing devices for patients with chronic obstructive pulmonary disease (Intermittent Positive Pressure Breathing Trial Group 1983), and drugs that stimulate heart contractions for patients with chronic heart failure (Packer, Carver, et al. 1991). In other cases, effective treatments are missed or delayed because the trials were not definitive. For example, the use beta-blocker drugs in chronic heart failure patients was delayed for several years, long after the benefits of these drugs in treating heart attack patients was known, because some believed that these drugs would be harmful to those with chronic heart failure. However, in the 1990s trials such as MERIT-HF (MERIT-HF Study Group 1999), CIBIS-II (CIBIS-II Investigators 1999), and CO-PERNICUS (Packer, Fowler, et al. 2002) established a 30 to 35 percent reduction in mortality in patients with mild to moderate chronic heart failure. In order to further investigate this issue of not relying on evidence-based medicine, consider the use of hormone replacement therapy in postmenopausal women as a case study.

Hormone Replacement Therapy and Cardiovascular Disease: A Case Study

The combination of estrogen and progestin is commonly used for successfully treating hot flashes and night sweats (vasomotor symptoms) in postmenopausal women. Many observational studies in postmenopausal women found lower rates of cardiovascular disease and osteoporosis in women who were taking estrogen than were observed in those not taking estrogen (Bush 1990; Stampfer and Colditz 1991; Grady et al. 1992; Psaty et al. 1994; Sidney, Petitti, and Quesenberry 1997; Bush et al. 1987; Henderson, Paganini-Hill, and Ross 1991; Newton et al. 1997). The consistency of this observation is notable. The lower risk of cardiovascular events was observed to be especially strong in women who had already experienced a previous cardiovascular event. Users of hormone replacement therapy (HRT) were observed to have 35 to 89 percent fewer recurrent events than non-HRT users. In addition, estrogen reduces serum lipid values and increases bone density (Rijpkema, van der Sanden, and Ruijs 1990; Cauley, Seeley, et al. 1995; Weiss et al. 1980; Cauley, Robbins, et al. 2003). Some studies suggested that adding progestin reduces the suspected risk of endometrial cancer in women taking estrogen alone (Writing Group for the PEPI Trial 1996). Based on this information, an estrogen-progestin combination as hormone replacement therapy in postmenopausal women became widespread (Roussow 1996; Keating et al. 1999).

However, this increase in the use of estrogen-progestin was not based on a randomized clinical trial with clinical outcome measures such as cardiovascular disease mortality and morbidity. In fact, in 1990 the commercial product for the estrogen-progestin treatment was presented to the U.S. Food and Drug Administration (FDA) for approval for the reduction of heart disease based on observational data and other laboratory results (Seaman 2003, 148–52). While the FDA advisory panel of external experts overwhelmingly voted for the approval of the commercial product despite concerns expressed by women's groups, the FDA did not approve the commercial product for this indication. The reason that the agency gave was the lack of randomized prospective trials demonstrating a reduction in heart disease outcomes with HRT and the design limitations of the existing observational data, which made interpretation challenging. This reduction in cardiovascular events could be the result of bias. For example, if women who chose to take HRT were healthier than those who did not, such an observation might occur. The FDA supported the development of prospective clinical trials to provide more definitive answers to this very important question about the role of HRT in women's health.

Two recent clinical trials have been conducted to investigate the role of hormone replacement therapy in postmenopausal women (Hulley et al. 1998; Writing Group for the WHI 2002). The results of these studies were contrary to what had been expected from the observational studies.

The Heart and Estrogen/progestin Replacement Study (HERS) was designed to test the hypothesis that use of HRT in a postmenopausal population of women with coronary disease would reduce the incidence of the combined events of nonfatal myocardial infarction or death from coronary heart disease (Keating et al. 1999). HERS was a randomized, double-blind, placebo-controlled trial of the daily use of HRT or a matching placebo. The trial began in early 1993 and by September 1994 had enrolled 2,763 postmenopausal women who had coronary disease and an intact uterus. The mean age of these women was 66.7 years. These women were followed for an average of 4.1 years. By the end of the first year of follow-up, more than 80 percent of the women were still taking their HRT medication. After three years, the compliance with HRT was 75 percent, a rate considered very good for trials of this size.

The HERS trial showed no significant difference in the rates of nonfatal myocardial infarction or death from coronary heart disease, even though HRT produced an 11 percent decrease in low density cholesterol levels and a 10 percent increase in the high density cholesterol levels when compared to placebo-treated women. Although there were no significant overall differences in the number of coronary heart disease events for the four years of follow-up (see table 15.1), the researchers did observe an increased risk during the first year of treatment for women on HRT. This increase was primarily caused by blood clotting related events. The almost doubling of the risk of thrombotic or clotting related events, in fact, was so striking that the investigators alerted the trial participants during the trial about the risks of blood clots (Grady, Hulley, and Furberg 1997).

These unexpected results of no difference in coronary events, despite reductions in cholesterol levels, and the increased risk of blood clots challenged the investigators to reconcile these findings with those of all the previous observational studies, common belief, and medical practice. The HERS population was somewhat older than the women in the observa-

Table 15.1 HERS Selected Major Results

Number of Subjects	E + P 1380 Events	Placebo 1383 Events	Percent Increase (+) or Decrease (−) in Risk
Outcome Treatment			
1. Primary Outcome			
Coronary Heart Disease	172	176	−1
CHD Death	71	58	+24
Nonfatal MI	116	129	−9
2. Death			
Total	131	123	+8
CHD	71	58	+24
Cancer	19	24	−20
3. Other Cardiovascular			
Stroke/TIA	108	98	+13
Coronary Bypass Surgery	88	101	−13
Percutaneous Coronary Revascularization	164	175	−5
4. Venous Thrombotic Event			
Total	34	12	18
Deep Vein Thrombosis	25	8	218
Pulmonary Embolism	11	4	179
5. Cancer			
Total	96	87	+12
Breast	32	25	+30
Endometrial	2	4	−51
6. Fracture			
Total	130	138	−5
Hip	12	11	+10

Notes:
CI=confidence interval; E + P =estrogen plus progestin; N = sample size; CHD=coronary heart disease; MI=myocardial infarction; TIA= transient ischemic attack

tional studies. And many of the earlier observational studies were of women using estrogen only instead of the estrogen-progestin combination used in HERS. The HERS population had previous coronary disease, whereas the observational studies were mostly in healthy women. While the HERS investigators were able to formulate no definitive explanations, they recommended that physicians not start HRT in women with coronary heart disease, although the researchers did not recommend that those already on HRT for a period of time stop taking the medication.

Among the conclusions reached by the HERS researchers was that the risk and lack of benefit accumulated for postmenopausal women with existing coronary disease. However, a much larger trial, the Women's Health Initiative (WHI), addressed the issue of HRT for postmenopausal women without preexisting coronary disease (Writing Group for the WHI 2002). WHI was a randomized, double-blind trial that evaluated interventions that might affect postmenopausal women's risk of heart disease, breast and colorectal cancer, and fractures. The three interventions were HRT, low-fat diet, and calcium supplementation. For the HRT portion, women were grouped according to whether they still had their uterus. Women with an intact uterus were randomized to receive either an estrogen-progestin combination or matching placebo. Women who had had their uterus removed were randomized to estrogen only or a matching placebo. In July 2002, the WHI published results for women with an intact uterus, comparing the estrogen-progestin combination with the placebo (Writing Group for the WHI 2002). The results were similar to those observed in HERS, again contrary to expectation. In this HRT arm of the trial, WHI randomized 16,608 postmenopausal women aged fifty to seventy-nine who had an intact uterus and no existing coronary disease with a planned follow-up of 8.5 years. The primary outcome was coronary disease, defined as nonfatal myocardial infarction or death resulting from coronary disease. Invasive breast cancer was the primary safety concern; that is, would use of HRT increase the risk of breast cancer. Other outcomes, such as total mortality, stroke, pulmonary embolism, endometrial cancer, colorectal cancer, and hip fracture, were also followed.

In May 2002, the Data and Safety Monitoring Board recommended that this component of the WHI be terminated early because of an excessive increased risk of invasive breast cancer (26 percent increase) and increased risk of other major clinical outcomes. These included unacceptable risk for coronary disease (29 percent increase), stroke (41 percent increase), and for pulmonary embolism (113 percent increase). There was no increased mortality risk. However, the researchers found that HRT offered benefits for those suffering from osteoporosis, although HRT did not affect the total mortality of these women. The authors concluded that the overall risks of estrogen-progestin exceeded the benefits, because such observed risks were unacceptable for a primary prevention intervention strategy. They recommended that estrogen-progestin not be started for coronary disease prevention or continued for postmenopausal women

Table 15.2   WHI Selected Major Results

Number of Subjects	E + P 8506 Events	Placebo 8102 Events	Percent Increase (+) or Decrease (−) in Risk
Outcome Treatment			
1. Cardiovascular Disease			
Total	694	546	+22
Coronary Heart Disease	164	122	+29
CHD death	33	26	+18
Non fatal MI	133	96	+32
Stroke	127	85	+40
Fatal	16	13	+20
Nonfatal	94	59	+50
Venous Thromboembolic	151	67	+111
Deep Vein	115	52	+107
Pulmonary Embolism	70	31	+113
2. Cancer			
Total	502	458	+3
Breast (Invasive)	166	124	+26
Endometrial	22	25	−17
3. Fractures			
Total	650	788	−24
Hips	44	62	−34
Vertebral	41	60	−34
4. Death	231	218	−2

Notes:
CI=confidence interval; E + P =estrogen plus progestin; N = sample size; CHD=coronary heart disease; MI=myocardial infarction; TIA= transient ischemic attack

with an intact uterus. More detailed results have now been presented for some of the individual outcomes assessed in the WHI (Cauley, Robbins, et al. 2003; Writing Group for the WHI 2002). The HRT component of the trial is continuing to evaluate estrogen-only therapy in postmenopausal women without a uterus.

While the relative risks of using estrogen-progestin are definitely increased, the absolute level of the risk is fortunately still very low. That is, the excess risks attributable to estrogen-progestin were seven additional coronary disease events, eight additional strokes, eight additional pulmonary embolisms, and eight additional cases of invasive breast cancer per 10,000 person years of exposure. The excess risk appears early primarily because of the clotting effects and then tends to level off, but early identification of who is at risk for blood clots is not now possible. However, women who need estrogen-progestin for treatment of postmenopausal vasomotor symptoms typically are a decade younger than those in the Women's Health Initiative. These women may choose to take HRT for a short period of time, perhaps postponing their symptoms until a later time, when, it is hoped, the symptoms will diminish. However, women should carefully weigh the benefits of short-term HRT therapy with the small but definite increase in risk (Grady 2003).

When the results of WHI are combined with the results of HERS, women with or without previous coronary disease who are taking estrogen-progestin appear to be at risk of clotting. Although the HERS trial was relatively small, WHI was a very large trial and provides a better estimate of the excess risk. In addition, a meta-analysis of all previous HRT trials came to similar conclusions about the risks and benefits (Bush 1990).

With the results of the HERS, WHI, and other meta-analyses, researchers have asked, "How could we have been so wrong?" (Nelson et al. 2002; Laine 2002; Humprey, Chan, and Sox 2002). The researchers say that one reason may be that the observational studies did not take socioeconomic factors into consideration. That is, higher socioeconomic status is associated with lower rates of cardiovascular disease and higher rates of HRT usage. The argument is that women who are healthy, active, and taking care of their other risk factors are also taking HRT, and it was not the use of HRT that kept these women healthy. However, associations between two factors do not infer causal direction; that is, which factor is cause and which factor is effect may not be clear, if any such association exists. Many other unknown factors could be responsible. However, there is no doubt that the consequences of not conducting early definitive clinical trials of HRT exposed millions of women to a therapy for reasons that now appear to be false. It should be noted that the FDA never approved the use of HRT for the prevention of cardiovascular disease, despite its widespread use for that purpose, because adequate data from well-controlled studies did not exist.

Lessons

We know that there are limits to what observational data and mechanistic theories can teach us (Taubes 1995). In some cases, observational data have been correct in identifying modifiable risk factors for disease. Undoubtedly, smoking increases the risk of lung cancer and cardiovascular disease, but even this point of view has its critics. The observational data indicating that high serum lipids and hypertension are independent risk factors for cardiovascular disease are also correct, although it took years to develop safe and effective interventions to modify those risk factors. However, many other potential risk factors suggested by observational data have not panned out (Taubes 1995). In some cases, the prior beliefs of clinicians and researchers about mechanisms of interventions have prevented or delayed advances in potentially effective therapies. This was certainly the case for the use of beta-blocker drugs in the treatment of chronic heart failure. The lesson of the WHI and HERS experience is that taking shortcuts in evaluating and promoting new interventions carries with it a greater chance of being incorrect. Some medications, procedures, and behavioral modifications are never properly tested before their use becomes widespread in practice. Medical devices often are tested using a surrogate or substitute for the ultimate clinical outcome, which is permitted by federal device regulations, and thus we cannot be sure in some cases that the device in fact has a clinical benefit. Since the early 1980s the medical research community has often been wrong because it did not follow through with definitive clinical trials, so these lessons are not new (DeMets and Califf 2002a, 746–51; DeMets and Califf 2002b, 880–86; Califf and DeMets 2002a, 1015–21; Califf and DeMets 2002b, 1172–75; Fleming and DeMets 1996). One problem is that there is not enough time, research teams, or funding to test every new therapy with rigorous clinical outcome studies. Such trials tend to be large, longer in patient follow-up, and more detailed in outcome measures and thus more costly. However, evidence-based medicine depends heavily on high-quality clinical trials with clinically relevant outcomes to make decisions about clinical and health policy (Tunis, Stryer, and Clancy 2003).

Health Policy and Practice Guidelines

Health policy depends not only on the risk-benefit ratio of a given intervention but also on evaluating alternative interventions and their cost-benefit ratios (Tunis, Stryer, and Clancy 2003). It is clear that we need more, not fewer, clinical trials in practice-based populations with clinical outcomes comparing the various treatment or intervention options currently available. Thus we will have to make choices and conduct more rigorous trials where the interventions have potentially serious side-effects or the disease has serious mortality or morbidity. For the interventions that we do not rigorously test, we will have to make our best judgments based on less than optimal evidence, always remembering what the strength of that evidence is. Sponsorship for the trials that we do conduct will likely have to come from not only federal research institutions such as the National Institutes of Health or private industry such as pharmaceutical or device manufacturers but also from health maintenance organizations (HMOs) and third-party payers such as Medicare. The latter organizations want to be sure they are paying for treatments and interventions that are safe, effective, and helpful. Not proceeding with more such trials means that more patients and healthy individuals will be exposed to ineffective therapies or even harmful interventions, perhaps at great personal and public financial cost. Evidence-based medicine is probably the best medicine, and we must strive to obtain the best evidence possible whenever it is feasible.

References

Bush, T. L. 1990. Noncontraceptive estrogen use and risk of cardiovascular disease: An overview and critique of the literature, pp. 211–23. In *The menopause: Biological and Clinical consequences of ovarian failure: Evolution and management,* edited by S. G. Korenman. Norwell, Mass.: Serono Symposia.

Bush, T. L. et al. 1987. Cardiovascular mortality and noncontraceptive use of estrogen in women: Results from the Lipid Research Clinics Program Follow-Up Study. *Circulation* 75:1102–9.

Califf, R. M., and D. L. DeMets. 2002a. Principles from clinical trials relevant to clinical practice: Part 1. *Circulation* 106:1015–21.

———. 2002b. Principles from clinical trials relevant to clinical practice: Part 2. *Circulation* 106:1172–75.

Cardiac Arrhythmia Suppression Trial Investigators. 1989. Preliminary report: Effect of encainide and flecainide on mortality in a randomized trial of arrhythmia suppression after myocardial infarction. *New England Journal of Medicine* 321 (6): 406–12.

Cauley, J. A., J. Robbins, et al. 2003. Effects of estrogen plus progestin on risk of fracture and bone mineral density. The Women's Health Initiative Randomized Trial. *Journal of the American Medical Association* 290 (13): 1729–38.

Cauley, J. A., D. G. Seeley, et al. 1995. Estrogen replacement therapy and fractures in older women. *Annals of Internal Medicine* 122:9–16.

CIBIS-II Investigators and Committees. 1999. The cardiac insufficiency Bisoprolol study II (CIBIS-II): A randomised trial. *Lancet* 353:9–13.

Connolly, H. M. et al. 1997. Valvular heart disease association with fenfluramine-phentermine. *New England Journal of Medicine* 337:581–88.

Coronary Drug Project Research Group. 1975. Clofibrate and niacin in coronary heart disease. *Journal of the American Medical Association* 231 (4): 360–81.

DeMets, D. L., and R. M. Califf. 2002a. Lessons learned from recent cardiovascular clinical trials: Part 1. *Circulation* 106:746–51.

———. 2002b. Lessons learned from recent cardiovascular clinical trials: Part 2. *Circulation* 106:880–86.

Diabetes Control and Complications Trial Research Group. 1993. The effect of intensive treatment of diabetes on the development and progression of long-term complications in insulin-dependent diabetes mellitus. *New England Journal of Medicine* 329 (14): 977–86.

Doll, R., and A. Hill. 1954. The mortality of doctors in relation to their smoking habits: A preliminary report. *British Medical Journal* 1: 1451–55.

Doll, R. et al. 1994. Mortality in relation to smoking: Forty years' observations on male British doctors. *British Medical Journal* 309 (6959): 901–11.

Early Breast Cancer Trialists' Collaborative Group. 1998. Tamoxifen for early breast cancer: An overview of the randomized trials. *Lancet* 351 (9114): 1451–67.

Fisher, B. et al. 1998. Tamoxifen for the prevention of breast cancer: Report of the National Surgical Adjuvant Breast Cancer and Bowel Project P-1 Study. *Journal of the National Cancer Institute* 90 (18): 1371–88.

Fleming, T. R., and D. L. DeMets. 1996. Surrogate endpoints in clinical trials: Are we being misled? *Annals of Internal Medicine* 125 (7): 605–13.

Friedman, L., C. Furberg, and D. L. DeMets. 1998. *Fundamentals of Clinical Trials.* 3d ed. New York: Springer-Verlag.

Grady, D. 2003. Postmenopausal hormones—Therapy for treatment only. *New England Journal of Medicine* 348 (19): 1835–37.

Grady, D. et al. 1992. Hormone therapy to prevent disease and prolong life in post-menopausal women. *Annals of Internal Medicine* 117:1006–37.

Grady, D., S. B. Hulley, and C. Furberg. 1997. Venous thromboembolic events associated with hormone replacement therapy. *Journal of the American Medical Association* 278:477.

Henderson, B. E., A. Paganini-Hill, and R. K. Ross. 1991. Decreased mortality in users of estrogen replacement therapy. *Archives of Internal Medicine* 151: 75–78.

Hulley, S. et al. 1998. Randomization trial of estrogen plus progestin for secondary prevention of coronary heart disease in postmenopausal women. *Journal of the American Medical Association* 280:605–13.

Humprey, L. L., B. K. S. Chan, and H. C. Sox. 2002. Postmenopausal hormone replacement therapy and the primary prevention of cardiovascular disease. *Annals of Internal Medicine* 137:273–84.

Hypertension Detection and Follow-up Program Cooperative Group. 1979. Five-year findings of the hypertension detection and follow-up program. 1. Reduction in mortality of persons with high blood pressure, including mild hypertension. *Journal of the American Medical Association* 242 (23): 2562–71.

Intermittent Positive Pressure Breathing Trial Group. 1983. Intermittent positive pressure breathing therapy of chronic obstructive pulmonary disease. *Annals of Internal Medicine* 99 (5): 612–20.

Kagan, A., T. Gordon, W. B. Kannel, and T. R. Dawber. 1959. Blood pressure and its relation to coronary heart disease in the Framingham Study. *Hypertension, Volume 7: Drug action, epidemiology, and hemodynamics.* Proceedings of the Council for High Blood Pressure Research, American Heart Association, New York.:

Kannel, W. B., W. P. Castelli, T. Gordon, and P. M. McNamara. 1971. Serum cholesterol, and the risk of coronary heart disease: The Framingham Study. *Annals of Internal Medicine* 74:1–12.

Keating, N. L. et al. 1999. Use of hormone replacement therapy by postmenopausal women in the United States. *Annals of Internal Medicine* 130:545–53.

Laine, C. 2002. Postmenopausal hormone replacement therapy: How could we have been so wrong? *Annals of Internal Medicine* 137 (4): 290.

MERIT-HF Study Group. 1999. Effect of metoprolol CR/XL in chronic heart failure: Metoprolol CR/XL randomised intervention trial in congestive heart failure. *Lancet* 353:2001–7.

Nelson, H. D. et al. 2002. Postmenopausal hormone replacement therapy: Scientific review. *Journal of the American Medical Association* 288 (7): 872–81.

Newton, K. M. et al. 1997. Estrogen replacement therapy and prognosis after first myocardial infarction. *American Journal of Epidemiology* 145:269–77.

Packer, M., J. R. Carver, et al. for the PROMISE Study Research Group. 1991. Effect of oral milrinone on mortality in severe chronic heart failure. *New England Journal of Medicine* 325 (21): 1468–75.

Packer, M., M. B. Fowler, et al. 2002. Effect of carvedilol on the morbidity of patients with severe chronic heart failure: Results of the Carvedilol Prospective Randomized Cumulative Survival (COPERNICUS) study. *Circulation* 106 17): 2194–99.

Peto, R. 1986. Influence of dose and duration of smoking on lung cancer rates. *International Agency for Research on Cancer Scientific Publications* 74:23–33.

Powles, T. et al. 1998. Interim analyses of the incidence of breast cancer in the Royal Marsden Hospital tamoxifen randomized chemoprevention trial. *Lancet* 352:98–101.

Psaty, B. M. et al. 1994. The risk of myocardial infarction associated with the combined use of estrogens and progestins in postmenopausal women. *Archives of Internal Medicine* 154:1333–39.

Rijpkema, A. H., A. A. van der Sanden, and A. H. Ruijs. 1990. Effects of postmenopausal estrogen-progesterone therapy on serum lipids and lipoproteins: A review. *Maturitas* 12:259–85.

Roussow, J. E. 1996. Estrogens for prevention of coronary heart disease: Putting the brakes on the bandwagon. *Circulation* 94:2982–85.

Scandinavian Simvastatin Survival Study Group. 1994. Randomized trial of cholesterol lowering in 4,444 patients with coronary heart disease: The Scandinavian Simvastatin Survival Study (4S). *Lancet* 344:1883–89.

Seaman, B. 2003. *The greatest experiment ever performed on women: Exploding the estrogen myth.* New York: Hyperion.

Shepherd, J. et al. 1995. Prevention of coronary heart disease with pravastatin in men with hypercholesterolemia. *New England Journal of Medicine* 333 (20): 1301–7.

Sidney, S., D. B. Petitti, and C. P. Quesenberry Jr. 1997. Myocardial infarction and the use of estrogen and estrogen-progestrogen in postmenopausal women. *Annals of Internal Medicine* 127:501–8.

Stampfer, M. J., and G. A. Colditz. 1991. Estrogen replacement therapy and coronary heart disease: A quantitative assessment of the epidemiologic evidence. *Preventive Medicine* 20:47–63.

Taubes, G. 1995. Epidemiology faces its limits. *Science* 269:164–69.

Tunis, S. R., D. B. Stryer, and C. M. Clancy. 2003. Practical clinical trials: Increasing the value of clinical research for decision making in clinical and health policy. *Journal of the American Medical Association* 290 (12): 1624–32.

U.S. Public Health Service. 1964. Smoking and health: Report of the Advisory Committee to the Surgeon General of the Public Health Service. Public Health Service Publication No. 1103. Washington, D.C.: U.S. Department of Health, Education and Welfare.

Veronesi, U. et al. 1998. Prevention of breast cancer with tamoxifen: Preliminary findings from the Italian randomized trial among hysterectomized women. Italian Tamoxifen Prevention Study. *Lancet* 352:93–97.

Weiss, N. S. et al. 1980. Decreased risk of fractures of hip and lower forearm with postmenopausal use of estrogen. *New England Journal of Medicine* 303:1195–98.

Writing Group for the PEPI Trial. 1996. Effects of hormone replacement therapy on endometrial histology in postmenopausal women: The Postmenopausal Estrogen/Progestin Interventions (PEPI) Trial. *Journal of the American Medical Association* 257:370–75.

Writing Group for the Women's Health Initiative Investigators. 2002. Risks and benefits of estrogen plus progestin in healthy postmenopausal women: Principal results from the Women's Health Initiative randomized controlled trial. *Journal of the American Medical Association* 288:321–33.

Wynder, E. L., and E. A. Graham. 1950. Tobacco smoking as a possible etiologic factor in bronchiogenic carcinoma: A study of six hundred and eighty-four proved cases. *Journal of the American Medical Association* 143(4): 329–36.

PART 4
Smallpox and Bioterrorism

16

Smallpox

The Disease, the Virus, and the Vaccine

Dixie D. Whitt

Throughout history smallpox killed untold millions of people and left even larger numbers of survivors disfigured for life. This scourge was finally eliminated from nature in 1977 by an effective worldwide vaccination program (Fenner 1982). This was not a minor undertaking. It involved a concerted effort not only by international health care workers and public health officials but also leaders from all nations, both developed and developing (Brundtland 2002). In some cases massive vaccination campaigns required the willingness of governments throughout the world to temporarily suspend hostilities in order to improve the health of their citizens.

If smallpox has been eliminated from the earth, why are we hearing so much about it? The answer, of course, is that although the disease has been eradicated from nature, the viruses that cause smallpox are still being maintained in certain laboratories in the United States and Russia in order to develop vaccines if the need ever arises. Those governments were supposed to have secured the stock cultures of the viruses, but rumors persist that other laboratories may also have strains. This raises a concern that the smallpox virus could be used as an agent of bioterrorism (Henderson et al. 1999). So now we have to think about how the civilized world can best prepare itself to deal with a smallpox outbreak if it were to occur.

Viruses are the smallest and simplest of the pathogenic microorganisms (Salyers and Whitt 2001). They can range from merely annoying (those that cause the common cold) to highly lethal (human immunodeficiency virus, the cause of AIDS). All viruses have a core, which contains a few genes and a few proteins that they need to initiate their replication cycle. This nucleoprotein core is protected by a capsid (shell) made up of tightly packed proteins. An additional layer, the envelope, surrounds the capsid of some viruses. Viruses differ considerably in shape and size, but even the largest ones are too small to be seen with a light microscope.

Smallpox is caused by the variola virus, a large (by virus standards), complex, brick-shaped virus that is covered by an envelope. Even complex viruses such as variola virus are very simple microorganisms that really cannot do much for themselves. Variola virus can infect only humans—an important factor that played a major role in the elimination of the disease. The virus does not live in other animals or free in the environment. Smallpox virus must live inside human cells and use the human cell's raw materials and machinery in order to carry out its few simple processes, which are basically to make more viruses.

The first step in any virus infection occurs when a surface protein of the virus recognizes a specific molecule (a receptor) on the surface of the target host cell and facilitates the attachment of the virus to the host cell (figure 16.1). Different viruses are transmitted in different ways, and they recognize receptor proteins that are located only on certain cells of the body. Host cells take in the attached virus, and the virus envelope and capsid are removed. This step, called uncoating, releases the nucleoprotein core, which is made up of the viral genome and a few associated proteins. The nucleoprotein core then directs the host cell to produce viral proteins and more copies of the viral genome (DNA in the case of variola virus). After many copies of the viral genome are made and viral proteins are produced, these components assemble to form new virus particles that are released from the host cell. Because each step is unique to a particular virus, each viral infection proceeds in its own way.

Smallpox has an incubation period of about twelve days (Breman and Henderson 2002). The infection starts when a person inhales the variola virus (figure 16.2). The virus attaches to the cells that line the upper respiratory tract. It then spends four to five days replicating in these respiratory cells. During this period the infected person is not contagious. At the end of this initial round of replication, the virus is shed into the saliva. This

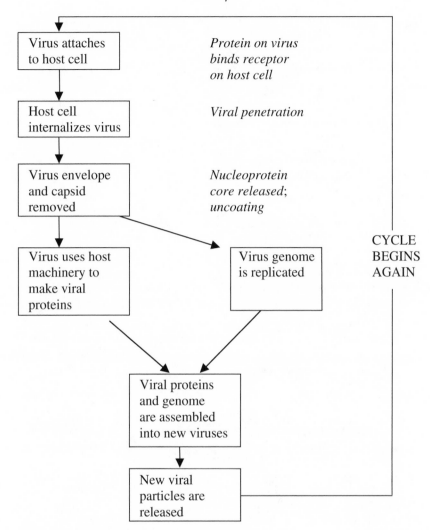

Figure 16.1 Viral infection of human cells.

means that the patient is technically infectious, but the number of viruses shed is very low, so transmission is unlikely to occur at this time. The viruses then enter the lymph nodes associated with the upper respiratory tract and move from there to the bloodstream. Once in the bloodstream they travel to the spleen, the bone marrow, and lymph nodes throughout the body where they undergo extensive replication in the cells of these tissues. In the lymph nodes they are picked up by white blood cells and once

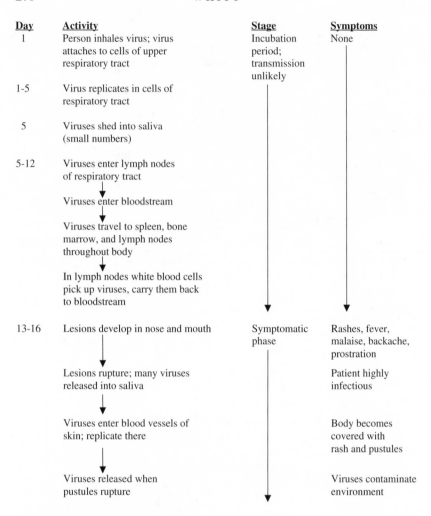

Day	Activity	Stage	Symptoms
1	Person inhales virus; virus attaches to cells of upper respiratory tract	Incubation period; transmission unlikely	None
1-5	Virus replicates in cells of respiratory tract		
5	Viruses shed into saliva (small numbers)		
5-12	Viruses enter lymph nodes of respiratory tract		
	Viruses enter bloodstream		
	Viruses travel to spleen, bone marrow, and lymph nodes throughout body		
	In lymph nodes white blood cells pick up viruses, carry them back to bloodstream		
13-16	Lesions develop in nose and mouth	Symptomatic phase	Rashes, fever, malaise, backache, prostration
	Lesions rupture; many viruses released into saliva		Patient highly infectious
	Viruses enter blood vessels of skin; replicate there		Body becomes covered with rash and pustules
	Viruses released when pustules rupture		Viruses contaminate environment

Figure 16.2   Progression of smallpox

again are carried to the bloodstream. This is the end of the incubation period, and the patient begins to suffer generalized flulike symptoms. Thus begins the symptomatic phase of the disease.

During the first three to four days of the symptomatic portion of the disease, the patient may suffer from rashes, fever, malaise, severe backache, and prostration. Early in this period lesions develop in the membranes of the nose and mouth. These lesions rupture quickly and release very large numbers of viruses into the saliva. This is the most infectious stage, and if the patient comes in close contact with others, the viruses can be spread by saliva droplets. Late in this period the characteristic "pox" lesions of the disease (the rash and pustules that cover the body) appear. This happens when the viruses enter the small blood vessels of the skin and multiply there. The patient is less likely to transmit the virus to others at this point because the concentration of virus particles in the saliva has decreased significantly, but contact with material from the pustules can result in infection.

The rash begins on the face and extremities but eventually covers the entire body. A day or two after the rash appears, it becomes blisterlike, and finally pustules develop. After a period of time the pustules rupture and crusts form. When the pustules rupture, they release viruses that can contaminate surfaces, clothing, and bed linen. Although many viruses are shed by this route, the patient is obviously very ill and is unable to go out in public. Usually, the only other people infected by such a patient are immediate caregivers. Large numbers of viruses are also present in the scabs that form, but they are not particularly infectious because they are bound tightly in the matrix of the scab. Once the scabs have come off, the patient can no longer transmit the virus to others.

Smallpox in the past was diagnosed by the appearance of the rash and pustules. At the beginning of an outbreak it was most often confused with chickenpox (Breman and Henderson 2002). Nowadays, if we were to be so unfortunate as to have a smallpox outbreak, advanced technology means the virus could be identified early on by any number of laboratory tests that would detect the specific variola DNA or proteins. Unfortunately, we still have no effective treatment against the smallpox virus, although scientists are testing antiviral compounds that have been developed since smallpox was declared a disease of the past. However, we do know how to contain the disease by isolating patients and vaccinating those who are at risk of being infected (World Health Organization 2001).

Vaccines work by convincing the body that it has been infected without actually having to endure the disease (Salyers and Whitt 2001). When the human body is confronted with a foreign agent, it makes an effort to defend itself by mounting an immune response that is specific to the invader. Many cells in the body are involved, and the process must be carefully choreographed. Vaccines may be living microorganisms that have been crippled so that they can no longer cause disease, or they may be dead microbes or components of microbes such as proteins or carbohydrates (subunit vaccines). A good vaccine can provide lifelong immunity to a particular microbe.

The smallpox vaccine was the first vaccine known to be used in humans. Its early history makes us shudder today, but when one considers what a dreaded disease smallpox was, it is not surprising that people were willing to tolerate an imperfect vaccine if it would prevent disease or lessen the severity of the disease.

Early cultures practiced various forms of immunization against smallpox. In the fifteenth century the Chinese developed an immunization method in which they covered bamboo splinters with the material from the pustules of patients with mild cases of smallpox, then introduced the material on the splinters into the nasal mucosa of others. This is called variolation because it used a weakened strain of the variola virus. The practice of variolation spread throughout parts of Asia and Africa and was finally brought to England from Turkey in the early eighteenth century by Lady Mary Wortley Montagu (Lowenthal 1994). Cotton Mather, the Puritan minister, introduced the procedure into North America in the 1720s. His action was so controversial that his house was firebombed. During the early part of the Revolutionary War about one-third of the newly recruited American troops died from smallpox, which had recently been introduced in Boston. Most British troops had become immune from earlier infections. Accordingly, George Washington made the decision to variolate all new recruits and quarantine them until they were no longer likely to transmit the disease to others. It had to be done in secret so that the British would not know that the American fighting force would no longer be dropping like flies, something the English had counted on in the past (Fenn 2001).

The disadvantage of variolation was that it used the variola virus. Many of those who were variolated developed the disease. Edward Jenner, an English country doctor, made a major improvement in the late eighteenth century (Fulginiti et al. 2003). He had observed that milkmaids, who often contracted a mild disease known as cowpox, were immune to smallpox. Jenner took the bold step in 1796 of inoculating a small boy with material taken from the pustular lesions of a cow infected with cowpox. Then, a few weeks later he tested the success of the procedure by inoculating the child with material from someone with smallpox. The child was immune. Today it is difficult for us to imagine how desperate the parents must have been to subject their child to this treatment. Such treatment became a hot political topic at the time. A political cartoon published in 1802 by the British satirist James Gillray shows vaccinated people with small cows growing out of their bodies. The procedure using cowpox became known as vaccination (*vacca* is Latin for cow).

Vaccination was a major advance over variolation because the virus in Jenner's vaccine did not contain variola but a close relative, cowpox, which was not likely to cause serious disease in those who were vaccinated. This is similar to the immunization procedure that is still used today, although we no longer use cowpox but another close relative, the vaccinia virus (Fulginiti et al. 2003). The vaccinia virus is also collected from infected calves.

These early vaccines against smallpox consisted of intact viruses, such as the weakened variola virus or the related cowpox virus. A later refinement was to use dead microbes as vaccines. These vaccines could induce an immune response yet not replicate in the cells of the inoculated person and cause disease. One problem with using dead microbes is that they do not elicit a particularly strong immune response. Living microorganisms will replicate in the body and elicit a more robust, long-lasting immune response because they provide a longer stimulation of the immune system. Because the vaccinia vaccine is a live virus vaccine, I will consider the development of an immune response to only that type of vaccine.

The immune system has two arms. One arm, cell-mediated immunity, relies heavily on cells called cytotoxic cells. The main activity of cytotoxic cells is to kill human cells infected with intracellular pathogens such as viruses. When human cells are infected by a virus, they often will display viral proteins on their surface. If a cytotoxic cell recognizes the particular viral protein, it will attack the infected cell. This attack involves the cytotoxic cell's binding to the infected human cell and then secreting toxic substances that kill the target cell. This is an effective strategy for controlling an infection, because by killing the infected cell, the cytotoxic cell prevents the replication of the virus and the release of many more viral particles capable of infecting many more target respiratory cells.

The second arm of the immune system consists of protein complexes called antibodies. Antibodies can bind to viruses and neutralize them. As I mentioned earlier, the first thing that a virus must do in order to cause an infection is bind to a specific receptor on the target cell of the host.

Antibodies are big bulky molecules; when they bind to a virus, they physically block the interaction between the virus and the human cell and thus prevent infection.

Because both arms of the immune response are tailor-made to deal with a specific microorganism, immunity is not going to happen instantaneously. This takes us back to my earlier statement that when the body develops an immune response specific to a new invader, the response requires many different types of cells and the careful regulation of their activities. This takes time.

For example, when a person is inoculated with vaccinia virus, the virus will encounter antigen-presenting cells (APCs), a type of white blood cell whose job is to pick up foreign intruders, degrade them, and present the degraded components on its surface. In the case of vaccinia virus, the degraded components would be parts of the proteins found in the envelope that surrounds the virus. As APCs circulate through the body, they will encounter other cells (T cells) that recognize the viral protein on the surface of the APC and bind to it. The number of T cells capable of recognizing a specific viral protein is very small—not enough to have any immediate effect on the large numbers of viruses being produced. However, binding of the T cell to the APC causes the APC to produce chemicals that stimulate the T cell to multiply and produce many identical T cells that will recognize the specific viral protein presented by the APC.

There are two kinds of T cells. Stimulation of one kind leads to the production of cytotoxic cells (which kill cells infected with the virus), while stimulation of the other kind leads to the production of antibodies (which bind to and neutralize the viruses circulating in the blood)—the components of the two arms of the immune system that I described earlier. It will take about a week to develop an adequate number of cytotoxic cells and antibodies to deal with the virus. Of course, during this period the virus is proliferating rapidly. In the case of the variola virus this is not good because the initial immune response cannot keep up with the viral replication, and the smallpox infection continues to develop in the patient. However, if we are dealing with vaccinia virus, replication of the virus is good because it continues to stimulate the immune response but does not cause disease. Thus the person inoculated with vaccinia virus (the vaccine) develops immunity not only to the vaccinia virus but also its close relative, the variola (smallpox) virus.

Another important aspect of the immune response is what is called immune memory. When an immune response is stimulated, most of the activity of the cytotoxic cells and antibodies is directed toward dealing with the current invader. However, a few memory cells are also produced. These cells remain in the body for a long period of time after the current infection is over. They are able and ready to multiply rapidly and respond quickly should the person ever encounter the same invader again. This is why preventative vaccination is so important. It primes the immune system, thus eliminating the lag time that occurs with a first exposure to either a vaccine or the pathogen.

At first it may seem somewhat surprising that not much is known about the role of the two arms of the immune system and the development of memory cells in regard to smallpox. When we consider that smallpox has been eradicated from nature, it is not all that surprising that scientists have not been doing much current research in this area (Fulginiti et al. 2003). In fact, in 1980, when smallpox was declared to be eradicated, T cells had just been discovered, and no one knew how they worked. It was much more productive to invest in research on the many diseases that were still killing millions of people every year than to research a disease that was thought no longer to be a threat to the world.

New studies are revealing that, indeed, both cytotoxic cells and antibodies are involved in immunity to variola (Hammarlund et al. 2003). And, very important, many of those who were immunized many years ago still have memory cells specific for the variola virus. As one would expect, the more "booster" immunizations a person had, the longer the immunity persists. These experiments have been done in the laboratory, and none of these subjects has been inoculated with variola to see if she or he is truly immune.

A lot of the controversy that we are hearing these days revolves around these issues. Are many middle-aged and elderly people still protected by their earlier vaccination? Do the adverse reactions to the vaccine outweigh the benefits that would occur if we were to have a bioterrorism event using variola virus? How would our public health departments respond to an outbreak of smallpox? Should everyone be vaccinated or just those exposed?

Those who advocate vaccinating just those people who have been exposed can point to the timetable of the disease's progression, and the timetable of immune response, to support their position. As figure 16.2 showed the incubation period for smallpox is prolonged. The viruses do not enter the bloodstream until at least Day 5 of the period, and the first lesions do not develop until about Day 13. If someone is exposed to small-pox and immunized within the first two or three days afterward, that person's immune response should develop adequately before Day 13. In earlier outbreaks it was shown that people who are vaccinated shortly after exposure either do not become infected or suffer a much milder case of the disease.

I hope that this brief overview of how the smallpox disease progresses from initial infection with the virus and how vaccination can stimulate the immune system to respond effectively against this virus will give readers a foundation for understanding the different views presented in the chapters that follow.

References

Breman, J. G., and D. A. Henderson. 2002. Diagnosis and management of small-pox. *New England Journal of Medicine* 346:1300–8.

Brundtland, G. H. 2002. Smallpox revisited. *Journal of the American Medical Association* 287:1104.

Fenn, E. A. 2001. *Pox Americana: The great smallpox epidemic of 1775–82.* New York: Hill and Wang.

Fenner, F. 1982. Global eradication of smallpox. *Reviews of Infectious Diseases* 4:916–30.

Fulginiti, V. A. et al. 2003. Smallpox vaccination: A review, Part 1. Background, vaccination technique, normal vaccination and revaccination, and expected normal reactions. *Clinical Infectious Diseases* 37:241–50.

Hammarlund, E. M. et al. 2003. Duration of antiviral immunity after smallpox vaccination. *Nature Medicine* 9:1131–37.

Henderson, D. A. et al. 1999. Smallpox as a biological weapon: Medical and public health management. *Journal of the American Medical Association* 281:2127–37.

Lowenthal, C. 1994. *Lady Mary Wortley Montagu and the eighteenth-century familiar letter.* Athens: University of Georgia Press.

Salyers, A. A., and D. D. Whitt. 2001. *Microbiology: diversity, disease, and the environment.* Bethesda, Md.: Fitzgerald Science Press.

World Health Organization. 2001. Smallpox. *Weekly Epidemiological Record* 76 (44): 337–44.

17

The Model State Emergency Health Powers Act

A Tool for Public Health Preparedness

Lesley Stone, Lawrence O. Gostin,
and James G. Hodge Jr.

Following the terrorist attacks of September 11, 2001, and the intentional release of anthrax spores that October, the Center for Law and the Public's Health at Georgetown and Johns Hopkins universities drafted model legislation designed to provide state and local public health agencies with legal authority to plan for, prevent, and respond to public health emergencies. The Model State Emergency Health Powers Act (MSEHPA) was intended to help states evaluate their existing public health laws, determine any areas that needed to be strengthened, and provide model language that could be used for improvement. (The model act is available at http://www.publichealthlaw.net/MSEHPA/MSEHPA2.pdf.) The MSEHPA was written in collaboration with the Centers for Disease Control and Prevention (CDC), National Governors Association, National Conference of State Legislatures, Association of State and Territorial Health Officials, National Association of County and City Health Officials, and the National Association of Attorneys General.

Since the second version of the model law was published on December 21, 2001, it has become one of the most widely used model public health laws in modern times. It has been introduced in whole or part in forty-three states, the District of Columbia, and the Northern Marianas Islands. Thirty-two states and the District of Columbia have enacted legislation based on provisions of the model law. Countless local governments have referred to it in considering their own legal powers (the measure may also apply to local public health agencies in some states). Government officials considering public health reforms in Australia, the United Kingdom, China, and Korea also have studied the model law.

Although widely used by law and policy makers, the MSEHPA has also been the subject of considerable debate and criticism. This chapter describes the need for a model law, discusses the main provisions of the MSEHPA, and responds to some of the most popular criticisms of the measure. While critics tend to focus on the individual rights that may be affected by the act, the act promotes the common good.

The Need for a Model Act

Part of the impetus for the MSEHPA was the existing status of state public health laws. The Institute of Medicine, in its 1988 and 2002 reports on the future of public health, noted that state public health laws in many jurisdictions were considerably outdated, obsolete, inconsistent, and fragmented (Gostin, Burris, and Lazzarini 1999, 102). In 2000 the U.S. Department of Health and Human Services noted in its report *Healthy People 2010* that "the Nation's public health infrastructure would be strengthened if jurisdictions had a model law and could use it regularly for improvements" (23-18). Many state public health laws do not conform to modern constitutional standards, adopt modern scientific principles, or provide clear legal bases for action (Gostin, Burris, and Lazzarini 1999). Federal and state public health authorities realized that an effective response to bioterrorism or other public health emergencies requires a strong legal basis and that model provisions could serve as an instructive template or checklist.

Before the MSEHPA was developed and used by state legislators, most public health statutes had not been systematically updated since the early to mid-twentieth century. Because the practice of public health is science-based, it is important that state laws be periodically reviewed to ensure that they are in accord with modern science. Similarly, conceptions of constitutionally required civil liberties have changed over the years, and the public health laws must be examined and amended to reflect the most recent rulings of the courts.

Without the widespread adoption of the MSEHPA, public health laws are inconsistent within states and among them (Gostin and Hodge 2002a; Gostin, Burris, and Lazzarini 1999, 102–3). Within states, these laws are fragmented, enacted decades ago to respond to major new diseases as they presented threats to the communal health. Because of this incremental approach to enacting public health legislation in many states, different rules apply to different diseases or conditions, often without any rational basis. For example, a New Jersey statute regarding communicable diseases (last updated in 1924) gives state and local boards of health the power to "maintain and enforce proper and sufficient quarantine wherever deemed necessary" (N.J. Stat. Ann. tit. 26, § 4-2 [West 1996]). However, New Jersey law also requires that public health authorities obtain a court order before they may isolate a person with typhoid fever (N.J. Stat. Ann. tit. 26, § 4-53 [West 1996]). There is no logical reason to accord those with typhoid fever more due process than those with other communicable diseases.

In addition, the structure, substance, and procedures for detecting, controlling, and preventing diseases vary profoundly between states (Gostin et al. 2002).[1] These inconsistencies could hamper prevention and response efforts for public health emergencies involving communicable diseases that affect multiple states. Because disease agents do not respect state borders, it is important that state laws either reach a basic level of conformity or require cross-border planning for emergencies. If a disease has spread across a state border, quarantining individuals in one state but not the other will not stop the spread of the disease and may cause an influx of exposed individuals into the state with more lenient quarantine rules. Efficient sharing of such resources as health care workers, medicine, and hospital supplies also depends on predetermined agreements between states. The MSEHPA was designed to make progress toward both objectives. It requires fundamental and comprehensive state public health planning for public health emergencies and recognizes the existing authority of states to enter into interstate compacts that are consistent with these plans. (Interstate planning is often achieved through emergency management assistance compacts between states.)

Many existing state public health laws fail to provide the authority necessary for detecting and responding to bioterrorism or naturally occurring infectious diseases. Some states have no provisions requiring planning or communication and coordination among the various levels of government (local, state, and federal), responsible governmental agencies, and the private sector. Yet effective planning, communication, training, and coordination between government branches and the private sector are critical to preventing or limiting the effects of a public health emergency. (Although states cannot compel the federal government to intervene, Congress has authorized federal cooperation with state and local authorities in many instances. See, for example, 42 U.S.C. § 243[a] [2004]).

Many states that have not adopted the MSEHPA do not require timely reporting of all biological agents identified as critical by the CDC (Gostin and Hodge 2002b). Similarly, states do not require, and some may prohibit, public health agency monitoring of data held by hospitals, managed care organizations, and pharmacies (Gostin and Hodge 2002a; Institute of Medicine 2000). State health departments may be limited in their ability to conduct syndromic surveillance (and other types of information gathering) that could alert authorities to unusual clusters of patients who are exhibiting potentially serious symptoms characteristic of biological agents. Because highly contagious diseases can spread exponentially, an effective monitoring system is needed at the state and local levels.

Even if a health threat is identified, public health authorities may lack legal authority to respond effectively under existing state laws. Public health authorities may need to exercise a variety of powers quickly, such as medical testing, screening, treatment, vaccination, quarantine, isolation, nuisance abatement, and taking of private property (e.g., medical supplies, drugs, vaccines, hospital beds). These authorities must make critical decisions about disease management for the communal good that are also consistent with civil liberties. The MSEHPA is intended to improve state public health law in two important ways. First, the act provides clear bases for exercising specific powers that state and local public health agencies may need for effective planning, detection, prevention, and response to a bioterrorist attack or other public health emergency. Second, the act modernizes existing law to reflect current scientific practices and constitutional protections for liberty and property interests. The MSEHPA has led states to analyze their public health preparedness and make decisions in an open forum about the best way to protect the public's health. The MSEHPA has been successful both as a catalyst for discussion and as a model law.

A Brief Description of the Provisions of the MSEHPA

Article I of the MSEHPA sets out the legislative findings about the need for the act, its purposes, and key definitions. Here, "public health emergency" is defined in relevant part as "an occurrence or imminent threat of an illness or health condition" that

(1) is believed to be caused by any of the following:
  (i) bioterrorism;
  (ii) the appearance of a novel or previously controlled or eradicated infectious agent or biological toxin; . . . and
(2) poses a high probability of any of the following harms:
  (i) a large number of deaths in the affected population;
  (ii) a large number of serious or long-term disabilities in the affected population; or
  (iii) widespread exposure to an infectious or toxic agent that poses a significant risk of substantial future harm to a large number of people in the affected population.

By imposing strict criteria, the definition is designed to trigger a public health emergency only in dire and severe circumstances.

The act also requires the state to plan for a public health emergency. Article II authorizes the governor to establish a commission to begin planning for such an emergency (§ 201). Within six months of the effective date of the measure, the commission must develop a comprehensive detection and response plan for public health emergencies. Thereafter, the plan must be reviewed and revised annually (§ 202). In addition, the measure provides for reporting and tracking threats to public health, as well as information sharing between public health and safety authorities (subject to privacy protections) when necessary to respond to serious public health threats (Article III).

Article IV prescribes the process of declaring such a public health emergency, from which additional governmental powers flow. The governor can declare a state of public health emergency (§ 401) by issuing an executive order. The declaration grants the governor a number of emergency powers, including the ability to (1) suspend the provisions of state statutes governing procedures for conducting state business if they would prevent necessary governmental action, (2) transfer the personnel or functions of state agencies, and (3) mobilize the militia (§ 403[a]). The declaration terminates automatically within thirty days. Additionally, a majority of the state legislature can terminate the declaration at any time (§ 405).

The declaration of a public health emergency also imbues public health authorities with certain powers. For example, during a declared public health emergency, public health authorities can close, evacuate, decontaminate, or destroy any facility or material that poses a danger to the public health (§ 501, § 507). Public health authorities are also able to control facilities and property in ways that are reasonable and necessary to respond to the emergency, dispose of infectious waste and human remains safely, and control health care supplies. In some cases, owners would be compensated for the use of their facilities or destruction of their property (Article V).

Public health authorities are also given powers to protect people during public health emergencies (Article VI). The public health authority is allowed to perform physical examinations and tests necessary to diagnose or treat individuals. People may refuse but could be isolated or quarantined if the authority is not able to determine that they do not pose a threat (§ 602, § 603). The act provides specific procedures for isolation and quarantine, including notice, hearings, and counsel for those affected by the implementation of such measures (§ 605). During a public health emergency, the public health authority must keep the public informed about health threats, in multiple languages if necessary (§ 701), and must provide access to mental health support personnel (§ 702).

Responses to Criticisms of the MSEHPA

Although the MSEHPA has generated significant legislative reforms at the state and local levels, considerable debates have surrounded its provisions. Here we respond to the most common forms of criticism: that the MSEHPA gives too much power and discretion to the state. In addition, the MSEHPA has been criticized as unnecessarily constricting the authority of public health agencies. Some critics would prefer that the agencies retain broad discretion to deal with public health emergencies (see Richards and Rathbun 2003).

These criticisms appear to have three main bases. First, critics view the MSEHPA within the context of the liberty-restricting climate that has prevailed since September 11, 2001. Civil libertarians aptly question the policies of the Bush administration's war on terrorism (ACLU 2002; Lynch 2002). Expansive powers in the areas of intelligence gathering and detention, such as military tribunals, restrictions on immigrants, and surveillance without judicial oversight, signal trends that liberty may be constricted in ways contrary to our traditions and without any real gain. During this same time period, the MSEHPA was released to a public that had not routinely reviewed government's traditional and inherent public health powers. The powers enumerated in the model legislation that related to managing property and protecting people during a public health emergency seemed to infringe unnecessarily on individual rights in an era of liberalism and strong observance of civil liberties. However, as we explain shortly, a careful reading of the model act reveals that such unchecked powers are not characteristics of the act. Instead, the measure sets forth mandatory criteria and procedural reviews that circumscribe government powers.

Second, many policy makers and the public were unaware that before September 11 many public health statutes granted broad powers that authorities could use to deprive citizens of liberty and property. These powers ranged from compulsory vaccination, treatment, and quarantine to destruction of property in the case of nuisance abatement. The laws granted health officials excessive discretion and made few provisions for due process. For example, Montana's state law on sexually transmitted diseases (STDs) gives health officers the power to isolate or quarantine those with an STD who refuse examination or treatment. The law further provides that "no one but the state or local health officer may terminate the isolation or quarantine." New Jersey law allows the public health agency to remove and destroy private property if, "in its opinion, [the property is] infected with any matter likely to communicate disease . . . when in its opinion the safety of the public health requires it." And California law allows public health authorities to quarantine or isolate "whenever in its judgment the action is necessary to protect or preserve the public health." In contrast, the MSEHPA, by specifying criteria for invoking emergency powers, reins in the potential for unfettered and unchecked discretionary acts of government officials. Similarly, the model act adds safeguards to curb the abuse of power by stipulating concrete procedural requirements. For example, the act requires a written petition to the court before isolating or imposing a quarantine on an individual, where possible, stipulates that the subjects of such an order are to have a hearing, and ensures that they are represented by counsel (§ 605).

Finally, the MSEHPA makes explicit the tradeoffs between individual liberties and common goods. By exposing the hard choices between personal rights and public security, the MSEHPA is a channel for the expression of competing societal viewpoints. Among the interest groups that have a stake in the outcome of the ensuing discussion are civil liberties organizations, privacy advocates, the health care industry, and the public health sector (including government public health agencies and nongovernmental public health interest groups). Because the model act provides a vehicle for discussing these difficult choices, it attracts criticism from multiple sources.

Specific criticisms focus on four main areas. Some critics advocate a federal, rather than state, solution to the problem. Others have expressed concern about the definition of a public health emergency. Economic and personal libertarians feel that the act gives too much power to government. And many have concerns about governmental abuse of power. We discuss these points in turn.

A Federal or State Solution?

Critics maintain that acts of terrorism are inherently federal matters, and federal authorities should determine the response (Annas 2002). Therefore, they suggest that updating state public health powers is not necessary. Although acts of international terrorism are undoubtedly also federal matters, states and their local governments are the traditional bulwark of public health in the United States (Gostin 2000). Historically, they have been the predominant actors in controlling communicable disease. Constitutionally, the states exercise the protection of public health under their police power (*Jacobson v. Massachusetts* [1905]). On a practical level, most public health activities take place at the state and local level. State government establishes the infrastructure for disease prevention, including state and local public health agencies, and the procedures for surveillance and protection.

In addition, states and localities will likely be the first to detect and respond to a public health emergency (Gostin 2003). Through established mechanisms, state health authorities can track incidence of disease that may signal that a bioterror attack has occurred. In other cases, emergency rooms may note an upsurge in uncommon illnesses, and in many states the law requires emergency rooms to relay this information to state or local public health authorities. It is impractical for the federal government to monitor such trends in all fifty states simultaneously. In a similar vein, states and localities are on the frontline of any response to a public health emergency. They have established local relationships and networks of health professionals and other actors in the public and private sectors whose roles during an emergency are critical.

Declaration of a Public Health Emergency

Critics are also concerned about the level of risk that may qualify as a public health emergency under the act (Annas 2002). They fear that a low-level risk would trigger a multitude of powers that the state could exercise. However, the act sets a high threshold for an emergency declaration.

As quoted earlier, the act defines a "public health emergency" as an occurrence or imminent threat of an illness that (1) is believed to be caused by either bioterrorism, the appearance of a novel or previously controlled or eradicated infectious agent or toxin (or potentially other causes as states may identify), and (2) poses a high probability of either a large number of deaths or a large number of serious or long-term disabilities in the affected population (MSEHPA § 104[m]).

These criteria are designed to ensure that only a true emergency can trigger the powers outlined in the model act. The agent must be one of bioterror, new, or previously controlled, and the risk to the population (in terms of number of deaths and long-term disabilities) must be significant. Contagious disease can spread exponentially,[2] and government action must be swift if it is to be effective. Some critics have claimed that the act could be used against people with HIV/AIDS or even influenza (Parmet and Mariner 2001). However, the MSEHPA does not support a formal declaration of a public health emergency for these conditions when existing public health measures provide for effective prevention and response. The act is structured to allow a public health agency to efficiently and responsibly mobilize around a rapidly emerging threat (unlike HIV/AIDS) to the population's health. Conditions such as normal strains of influenza do not trigger an emergency declaration because they are not novel and do not pose a high probability of a large number of deaths or disability in the population (Thomson et al. 2003).

Personal Libertarianism

Some claim that the government should not have compulsory powers over people for vaccination, testing, treatment, quarantine, and isolation (Annas 2003). Fundamentally, they contend that individuals will comply voluntarily with public health advice, and thus tradeoffs between civil rights and public health are not required. Yet life is full of examples of individuals who fail to comply with government orders and medical advice. One Harvard Medical School professor estimates that "as much as 50% of medications prescribed for chronic use never get taken" (Wiebe 1999, 27). There is always the potential that an individual or group will not comply with an order that restricts freedom.

The MSEHPA supports and anticipates high rates of voluntary compliance with public health measures. As Leavitt's tale of smallpox in chapter 19 demonstrates, voluntary compliance among the majority is essential if disease control measures are to be effective. Strong communication between the government and the population is necessary to build the trust that makes residents more likely to follow reasonable orders. Indeed, chapter 18 explains the failure of the recent smallpox vaccination campaign as partially the result of the government's failure to make the case for inoculation. Well-structured legal provisions are merely part of what may be required to ensure appropriate public health responses to a bioterrorism attack or other public health emergency.

However, at some point, to protect the larger population, government must have the power to coerce those who do not comply voluntarily. Liberal scholars acknowledge the "harm principle," that government should have the power to prevent individuals from endangering others, even when this interferes extensively with the individual's autonomy or other protected interests (Mill 1859; Feinberg 1984). This principle (and supporting legal theories) suggests that a person with a contagious, deadly disease has no "right" to engage in behavior that is likely to infect others. The MSEHPA thus provides mechanisms for restraining those individuals that would harm others through noncompliance, while requiring strong adherence to due process protections and other constitutional norms.

Economic Libertarianism

Critics suggest that businesses and health care professionals should not be subjected to burdensome regulation, even during a public health emergency (Lueck 2002). Yet businesses must comply with a variety of health and safety regulations that encroach upon economic freedom. Regulations requiring regular inspections and compliance with complex licensing schemes restrict business owners for the common good. Those who believe in the absolute power of the market may not agree with such regulations, but they are necessary to ensure that business activities do not endanger the public.

In a similar vein, government has always had the power to confiscate private property for the public good under its powers of eminent domain. Property may be subject to governmental takings, in which case the owner must be paid fair compensation (*Lucas v. South Carolina Coastal Council* [1992]). In addition, the government's authority to abate a nuisance is traditionally well accepted (see, for example, *Board of Health of City of Yonkers v. Copcutt* [1893]; *Ex Parte* Hughes [1939]; and *People v. Wheeler* [1973]). During an emergency involving a biological agent, the state will need to have adequate supplies of vaccines or pharmaceuticals and to have the use of health care facilities for medical treatment or quarantine. Without statutory provisions that clearly delineate these powers, access to these products or facilities could be delayed or prevented.

Government Abuse of Power

Critics warn that governors and public health authorities can abuse their powers under the MSEHPA (Parde 2002). First, it is important to understand that many public health laws before 2001 granted public health authorities even broader discretion and power to deal with public health emergencies. "A traditional state law establishing a health department might say little more than that the department had all necessary powers to protect the public health," according to Richards and Rathburn (2003). The MSEHPA is an improvement over such laws because it limits this discretion in several important ways.

The MSEHPA constrains government overreaching by providing clear criteria and a system of checks and balances. The governor may declare an emergency only after several criteria have been met and careful consultation with public health experts and the community (MSEHPA § 401). In addition, the legislature can check the governor's ability to declare such an emergency by overriding the governor's declaration at any time by majority vote (MSEHPA § 405). The power of the executive branch is also checked by the judiciary, which can terminate the exercise of power if the governor acts in ways that are not in accordance with the act's standards and procedures. The MSEHPA thus incorporates greater safeguards than many public health laws in existence before it. In addition, the model act signals the importance of individual liberty and respect for individuals in its preamble and throughout. The measure reflects the principles of significant risk, least restrictive alternative, and humane care and treatment of all people, reinforcing these values in our democracy.

Conclusion

The challenge for governments in responding to terrorism lies in striking the proper balance between protecting the public's health and safeguarding individual rights. A considered response requires planning and must allow for government to act swiftly to keep disease from spreading while protecting individuals from unnecessary government interference. The Model State Emergency Health Powers Act was designed to construct such a balance. It has provided model language for many states and promulgated public discussion of individual liberties and the common good.

Many criticisms of the MSEHPA, while adding important perspectives to the discussion, do not hold up under scrutiny. While critics argue that bioterrorism requires a federal response, strong state legislation is crucial to bioterrorism preparedness. In the model act, the declaration of a public health emergency is designed to prevent overextension of governmental powers by requiring specific conditions as prerequisites to such a declaration. Procedural safeguards protect personal and economic liberties but do not supersede the common good. Finally, both a system of checks and balances and procedural requirements curb government abuse of power.

Notes

This commentary is based on Lawrence O. Gostin, "The Model State Emergency Health Powers Act: Public Health and Civil Liberties in a Time of Terrorism," *Health Matrix: Journal of Law-Medicine* 13 (2003): 3–32; and Lawrence O. Gostin, "Public Health Law in an Age of Terrorism: Re-thinking Individual Rights and Common Goods," *Health Affairs* 21 (2002): 79.

1. Until recently, no uniform or model public health code had ever existed to promote interstate consistency (Gostin, Burris, and Lazzarini 1999, 102). On Sep-

tember 16, 2003, such a model act was published by the Turning Point project, an initiative of the Robert Wood Johnson Foundation and the W. K. Kellogg Foundation. The Turning Point Model State Public Health Act is available at http://www.turningpointprogram.org/Pages/MSPHAfinal.pdf. (accessed July 20, 2004).

2. For example, following a 2001 simulation of a smallpox attack, called "Dark Winter" and run by the Center for Strategic and International Studies, the Analytic Services Corporation, the Johns Hopkins Center for Civilian Biodefense Strategies, and the Oklahoma National Institute for the Prevention of Terrorism, statisticians estimated that more than three million Americans could have been infected during a two-month period, with more than one million deaths. The theoretical source of this outbreak was a group of terrorists who had infected themselves with the smallpox virus (Danzig 2003). For this reason, "even as few as 50–100 cases would likely generate widespread concern or panic and a need to invoke large-scale, perhaps national emergency control measures," according to the Center for Civilian Biodefense Strategies(http://www.hopkins-biodefense.org/pages/agents/agentsmallpox.html [accessed September 30, 2003]).

References

ACLU. 2002. Insatiable appetite: The government's demand for new and unnecessary powers after September 11. An ACLU report, New York.

Annas, G. J. 2002. Bioterrorism, public health, and civil liberties. *New England Journal of Medicine* 346 (April 25): 1337.

———. 2003. Terrorism and human rights. In *In the Wake of Terror: Medicine and Morality in a time of crisis*, edited by J. D. Moreno. Cambridge, Mass.: MIT Press.

Board of Health of City of Yonkers v. Copcutt, 35 N.E. 443 (N.Y. 1893).

Danzig, R. 2003. Academics and bioterrorists: New thinking about the new terror, *Cardozo Law Review* 24:1497.

Ex Parte Hughes, 129 S.W.2d 270 (Tex. 1939).

Feinberg, J. 1984. *The moral limits of the criminal law*. Vol. 1: *Harm to others*. Oxford: Oxford University Press.

Gostin, L. O. 2000. *Public health law: Power, duty, restraint*. Berkeley: University of California Press.

———. 2003 When terrorism threatens health: How far are limitations on personal and economic liberties justified? *Administration and Regulatory News* 28:6.

Gostin, L. O., and J. G. Hodge. 2002a. The model health emergency powers act: Why is it important now? *Northwest Public Health*, pp. 16–17.

———. 2002b. *Model state emergency health powers act commentary*. Seattle: Turning Point National Program Office, University of Washington.

Gostin, L. O., S. Burris, and Z. Lazzarini. 1999. The law and the public's health: A study of infectious disease law in the United States. *Columbia Law Review* 99:59.

Gostin, L. O. et al. 2002. The Model State Emergency Health Powers Act: Planning for and response to bioterrorism and naturally occurring infectious diseases. *Journal of the American Medical Association.* 288 (5): 622–28.

Institute of Medicine. 1988. *The future of public health.* Washington, D.C.: National Academy Press.

———. 2000. *Ending Neglect: The elimination of tuberculosis in the United States.* Washington, D.C.: National Academy Press.

———. 2002. *The future of the public's health in the twenty-first century.* Washington, D.C.: National Academies Press.

Jacobson v. Massachusetts, 197 U.S. 11 (1905).*Lucas v. South Carolina Coastal Council,* 505 U.S. 1003 (1992).

Lueck, S. 2002. States seek to strengthen emergency powers: Movement is raising privacy and civil-liberties concerns. *Wall Street Journal,* January 7.

Lynch, T. 2002. Breaking the vicious cycle: Preserving our liberties while fighting terrorism. Policy analysis, CATO Institute, Washington, D.C.

Mill, J. S. 1859, 1978. *On liberty.* Edited by Elizabeth Rapaport. Indianapolis, Ind.: Hackett.

Parde, D. 2002. CDC proposal is extreme. *USA Today,* April 25.

Parmet, Wendy, and Wendy Mariner. 2001. A health act that jeopardizes public health. *Boston Globe,* December 1, p. A1.

People v. Wheeler, 30 Cal. App. 3d 282 (1st Dist. 1973).

Richards, E. P., and K. C. Rathbun. 2003. Review of the Model State Emergency Health Powers Act. Louisiana State University Program in Law, Science, and Public Health White Paper #2, April 21, 2003. http://biotech.law.lsu.edu/blaw/bt/MSEHPA_review.htm (accessed June 21, 2004).

Thomson, W. W. et al. 2003. Mortality associated with influenza and respiratory syncytial virus in the United States. *Journal of the American Medical Association* 289 (2): 179–86.

U.S. Department of Health and Human Services. 2000. *Healthy people 2010.* Washington, D.C.: U.S. Department of Health and Human Services.

Wiebe, C. 1999. Following orders: New solutions to the age-old problem of patient noncompliance with prescription drug orders. *American Medical News* 42 (December 20): 27–30.

18

The States and the War against Bioterrorism

Reactions to the Federal Smallpox Campaign and the Model State Emergency Health Powers Act

David Rosner and Gerald Markowitz

Introduction

The newspaper headlines were stark and eerie: "Efforts to Calm the Nation's Fears Spin Out of Control," "Local Public Health Officials Seek Help," "This Is Not a Test," "Some States Can't Handle Bioterrorism," "Scared into Action." And the pictures that accompanied them were worse: space-suited investigators, smallpox-ridden children, cold stark laboratories staffed by masked personnel. State and local health departments were now supposed to be on a "war footing," as one headline noted.

The events of September 11 and the October 2001 anthrax incidents placed public health and public health agencies in the spotlight to a degree not experienced since the great epidemics of influenza, polio, whooping cough, and the like during the first fifty years of the last century. Many officials felt overwhelmed. The limitations of the public health surveillance system, laboratories, and treatment and social services became all too apparent. Beleaguered staff, limited laboratory space and supplies, and the general inexperience with bioterrorism led to a profound reevaluation—sometimes naive, sometimes quite sophisticated—of the place of population health services in the country's antiterrorism and emergency preparedness systems.

Two years after the attack, many public health officials believed there was still tremendous ambiguity about what bioterrorism and emergency preparedness really mean. Some saw emergency preparedness as synonymous with strengthening the existing public health infrastructure. Some saw it as building population health services more broadly. Others saw it as narrowly focused on smallpox, anthrax, emergency care, border protection, and the like. While all these formulations are obviously complementary, they often create competing demands for scarce resources.

This chapter is a contemporary history of two important initiatives that affected the various states' public health systems in the two years after the September 11 terrorist attacks: the smallpox inoculation campaign and the circulation of the draft Model State Emergency Health Powers Act. As historians, we hope to provide the lessons learned and perspectives gleaned by the participants themselves, using a wide variety of primary sources, including published and unpublished reports, oral interviews with key participants, and government documents and popular media. Certain themes emerged from the interviews, and the primary and secondary sources that we examined: In the immediate weeks after September 11, there was widespread hope among the professionals involved that the new focus on public health would result in a revitalization of the field and dramatic improvements in the public health infrastructure; states soon faced a conflict between widespread budgetary problems and the federal mandate in late 2002 for a major smallpox inoculation campaign; and the attempt to codify and reformulate state public health laws through the distribution of the draft Model State Emergency Health Powers Act, combined with what some criticized as the poor federal handling of the anthrax episode and smallpox inoculation campaign, stimulated a broad discussion of the obligations and responsibilities of health authorities in light of the new geopolitical situation. In general, our interviews with key participants and our reading of published and unpublished reports and media coverage suggest that the early potential to reform the public health system is endangered. Those services are essential for the improvement of the nation's efforts to address bioterrorism preparedness and the overall health needs of the population.

Smallpox Campaign

Once a disease that swept through cities and towns, leaving in its wake untold numbers of casualties and disfigured people, smallpox has disappeared from the natural environment through massive inoculation and public health campaigns worldwide. Yet for decades, both the United States and the Soviet Union stored the virus that causes smallpox, distributed it to various research labs, and made other moves that ensured that this public health victory became a potential human tragedy. Not only did the two countries refuse to destroy the remaining vials of smallpox virus, but evidence exists that some scientists in both countries, working at the behest and direction of their governments, labored hard to figure out how to make it all the more deadly and immune to the very public health measures that had "eradicated" it in the first place. By genetically altering the virus and selecting out strains that will not be intimidated by the smallpox vaccines of earlier vaccination campaigns, cold warriors may have effectively created a world in which we can all worry.

This was the irony of the call to vaccinate hundreds of thousands of emergency workers and to achieve "voluntary" compliance by all the other residents of the United States: The vaccines developed in U.S. labs may be effective against older strains of the virus and perhaps the newer, more powerful strains that we have developed, but it is unclear whether they would protect against strains purposely modified or developed in the old Soviet Union. Furthermore, while federal officials focused on the one biological agent that they believed that they might control, it was not clear that mass campaigns to inoculate the population wouldn't kill more people than would exposure to the virus itself. In 1947, the last time a mass campaign to inoculate against smallpox was carried out in New York City, three people died of smallpox while perhaps another dozen died from the aftereffects of the vaccination itself.

Despite the profound ambiguities regarding the dangers from smallpox, the events of September 11 and the anthrax episode sent the federal government into panic mode, as it tried to show that it was in control of events that were not controllable. The federal government mandated that states begin to prepare emergency response plans and to identify weaknesses in their systems. Part of this mandate was to prepare first responders capable of functioning in case of a smallpox attack.

Initially, this national attention to public health and its needs was seen as an enormous boon to the field. Public health officials saw the smallpox campaign as a win-win situation: Their departments could play an important role in the antiterrorism efforts while focusing the nation on the importance of public health activities in general. Further, federal money could save thousands of positions in state departments of health where state budgets were being cut as legislators worried about growing deficits. From California to Maine, state officials initially saw in the federal mandate a possibility for saving scores of public sector jobs, shuffling job titles so that workers in well-baby and childhood inoculation campaigns could be hired with newly available federal funds. From the first, states mobilized to organize broad inoculation campaigns involving state and local health care workers.

If improved communication and coordination among public health and health care entities for planning and conducting smallpox vaccinations count, the program has some success stories. In fact, the debate itself had a positive effect. J. Nick Baird, director of the Ohio Department of Health, put it this way: "In an open society we have to expect differences of opinion." But the smallpox debates helped "the nation move quickly to address some ascendant concerns and will put us in good stead should we need to move quickly on an outbreak in the future," he told us. Despite the questionable goal of the federal government to inoculate a half-million people, in fact, the public health community "essentially decided at what level we should vaccinate for smallpox. Although the numbers may be smaller than many states desired," Baird notes, "we will have a cadre of persons who can respond in the event of need."

Nevertheless, despite the early optimism and planning, many, if not most, of the states began to encounter resistance to the campaign. The smallpox vaccination campaign, initiated in December 2002, added to the strains on state public health departments, which were facing severe budget crises. Support began to wane as the vaccination campaign continued and it became clear that this effort—far from building support for and strengthening population health initiatives in the states— was actually draining resources and energies. Although some officials embraced the opportunity to plan for bioterrorist events, soon an uncommitted public health community, a resistant public, and health providers who did not volunteer to become vaccinated virtually stopped the campaign in its tracks. As of June 2003, the goal of vaccinating half a million people was far from being met. Only 37,608 health care workers had received the vaccine, according to reports presented at a conference called by the Centers for Disease Control and Prevention (McNeil 2003). In California, for example, the initial goal of the campaign was vaccinating forty thousand but, according to Angela Coron, associate director of the California Department of Health Services, the state had inoculated only "somewhat over a thousand" by the end of March (Coron 2003).

One major issue in California and other states was the safety of the vaccine itself, although the smallpox vaccine cannot cause the disease because it contains vaccinia, not the smallpox virus. No systematic smallpox vaccination campaign had been conducted for a number of decades, and the relative risk of the vaccine seemed to many to outweigh the potential advantages in the event of an attack.

In Minnesota the state quickly "ran into some trouble with some of our doctors," said state representative Lee Greenfield, as evidence surfaced that the vaccine could create problems for people with heart disease. Ohio health officials faced concern about potential heart problems as well as worries "about spreading the live [cowpox] virus to AIDS patients," who might be particularly vulnerable because their immune systems were compromised, according to Anne Harnish, assistant director of the state health department. Infectious-disease doctors were concerned that the vaccine might lead to outbreaks of smallpox as patients shed the virus, even though smallpox vaccine contains no live or even smallpox virus, she said. In Texas the state epidemiologist, Dennis Perrotta, the architect of the program there, had "grave concerns about inoculating large numbers of people." When a local reporter asked whether he would be vaccinated, Perotta replied, "I will be immunized and I will live someplace else for three weeks (while the vaccination site sheds virus)" (Roser 2003).

In addition to safety concerns, many public health personnel, embroiled in horrific budget fights, were concerned about the cost of the campaign. If state officials were convinced a threat was imminent, the response might have been different. But in Maine, at least, such was not the case. According to Charlene Rydell, health policy adviser to U.S. Rep. Tom Allen, D-Maine, "There is not a strong enough threat at this time, especially when we're having budget problems, and we're getting resistance from individuals especially with publicity about health problems and deaths. The federal government has not come through with funds for the true costs of the program, which costs about two hundred dollars per person when there are no complications."

Health care professionals and hospitals had particular concerns that led to their unwillingness to participate in the program. Many institutions were apprehensive about liability issues, particularly whether the institution would be held liable if someone suffered disease or death as a result of being inoculated, or patients became ill as a result of being exposed to hospital staff who had been exposed to those recently vaccinated. For staff members, the issue was whether they would be covered by workers' compensation if they became ill as a result of being inoculated. In several states, a number of hospitals "opted out" because of these fears, said Lee Greenfield, chair of the Minnesota House Finance Division of Health. In some states, hospital officials feared that their institution might be designated the emergency smallpox hospital for the area. Such a designation could financially ruin an institution because patients would go elsewhere for other health treatment for fear of being exposed to smallpox.

The uneasiness that arose while considering the safety of the vaccine and the financial costs of the vaccine were concrete manifestations, some informants asserted, of a growing discomfort with the entire rationale for the smallpox campaign. As Greenfield summarized that perspective, "There are a lot of people now reluctant . . . to get immunized. There is the whole public health question that people have raised. We do not know of an actual case of smallpox that exists in the world." Thus, he asked, "Why are we doing this?" Many public health administrators throughout the country had a general sense that a political agenda had trumped public health judgment. Many voiced the sentiment that public health should be left to public health professionals, rather than risk inaccurate public messages that would undermine public health authority.

The three-pronged program to inoculate the nation's military, emergency responders, and, ultimately, the general population against smallpox had a profound effect on public health officials and agencies around the country. The destructive redirection of scarce resources during times of severe budgetary constraints at the federal and state levels doomed this program to an eventual silent death. In the meantime, the mundane activities of the public health community—making sure that the water is safe to drink, babies are immunized, the air isn't too polluted to breathe, restaurants maintain sanitary practices, and supermarkets don't sell spoiled food—were all too often ignored. Across the country, state administrators referred to the distortions in program and policy that the smallpox campaign had caused. In Virginia, for example, county "departments of health have had to give up some of their core functions," according to Lisa Kaplowitz, deputy commissioner for Emergency Preparedness and Response for the Virginia Department of Health.

A number of state administrators saw the federal government's insistence on a smallpox campaign as, at a minimum, an intrusion, if not outright pressure. As Kaplowitz reported, "To say that there was pressure from the federal government was an understatement. . . . They say they are not mandating it, but the pressure is intense." In the end it was a "security issue," not a public health issue. But even as a security issue, the focus on smallpox was regarded by state officials as taking time and resources from other pressing bioterrorist threats. "Somebody high up believes there is some likelihood that smallpox is a terrorist issue. The downside is that you're not looking out for agents that are more likely. There are a lot of other organisms that are available to many people—plague, anthrax, botulism, et cetera," Kaplowitz said. "We are not focusing on the broader issues of biological, chemical, or radiological terrorism. They're being overshadowed by smallpox."

The Draft Model Act and the States

In the aftermath of September 11, conflicting values concerning the rights of individuals and the perceived needs for greater bioterrorism preparedness heightened the sense of disorder for officials whose jobs had been radically transformed in a very short time. Many public health practitioners and policy makers in the early twenty-first century are confused about their powers, and even their responsibilities, with regard to infectious disease control. For much of the nineteenth century, as epidemic disease swept through the nation's growing cities and was the major cause of death for children and adults alike, few public health officials questioned their right to engage in activities that intruded on personal liberties and civil rights. But as epidemic diseases waned in the twentieth century as the major causes of death, and as preventative measures and curative techniques developed, the drastic use of forced quarantine, isolation, and involuntary removal to infectious disease hospitals virtually ceased as everyday tools of public officials. With the very limited exceptions of AIDS and tuberculosis during the 1980s, civil liberties concerns outweighed the police powers of public health officials. As Lawrence Gostin has argued, embodied in the Constitution is an inherent tension between "the legal powers and duties of the state to assure the conditions for people to be healthy . . . and the limitations of the power of the state to constrain the autonomy, privacy, liberty, propriety, or other legally protected interests of individuals for the protection or promotion of community health" (Gostin 2000, 4; see also Salinsky 2003, 12).

In this context, the distribution of the draft Model State Emergency Health Powers Act, an attempt to codify and reformulate state public health laws, stimulated a broad discussion of the obligations and responsibilities of health authorities in light of the new geopolitical situation. The model law became a lightning rod in the debate about the implications for public health law of bioterrorism and chemical attack. It focused many public health officials on serious questions regarding the rights of federal, state, and local health officials to organize quarantine, isolation, surveillance activities, and the like and their potential infringement on civil liberties.

At the behest of Gene Matthews, legal counsel to the Centers for Disease Control and Prevention, Stephen Teret and Lawrence Gostin at the Center for Law and the Public's Health at George Washington and Johns Hopkins universities, drafted a piece of model legislation, called the Model State Emergency Health Powers Act (MSEHPA), with staff of the National Conference of State Legislatures, the National Governors Association, the National Association of Attorneys Generals, the Association of State and Territorial Health Officials, and the National Association of County and City Health Officials. Although this effort began well before September 11, 2001, the attack on the World Trade Center and the Pentagon, coupled with the anthrax outbreaks a month later, greatly increased the urgency of the enterprise. By the end of October 2001, a draft model act had been posted on the Internet and was circulating throughout the country (Gostin et al. 2002; Gostin 2002; Colmers and Fox 2003).

The model act was conceived as a checklist, or template, that the states could use in assessing their existing emergency public health laws. The model act itself was not the first attempt to recodify existing public health law. But, unlike earlier efforts, the stress of the moment lent the effort urgency. The model act would grant broad authority to governors to declare a state of emergency in the event of a bioterrorist attack. It would further allow public health officials to gain access to personal health records without the usual patient consent. It would require physicians and pharmacists to report "unusual" health events and to require physicians to provide information regarding individual patients who show unusual symptoms. The act would also allow public health officials to initiate quarantine and isolation measures and to mandate vaccinations and medical examinations. Public health officials would also be given the power to "seize and control" personal property and access to communications. Finally, it would grant to governors exclusive power over "funds appropriated for emergencies" and require states to develop comprehensive plans for responding to attack (Bourne and King 2001).

From across the political spectrum, "there appeared to be no great enthusiasm for the gradual recodification foreseen when the act was first released" (Bayer and Colgrove 2003, 65). Even in the aftermath of the September 11 attacks, some perceived the provisions of the model act as a "governmental takeover" (64). Organizations such as the National Council of State Legislatures, which ultimately endorsed the model act, found that some of their members had "gone ballistic" when they learned that they were "being listed as a collaborator on the draft" (Bayer and Colgrove 2003, 64). As reported in the national media, opposition to the model law appeared to be intense. Groups from the left to the right highlighted the possibilities of government intrusion into the lives of ordinary Americans. In a front-page article in *USA Today* headlined "Many States Reject Bioterrorism Law; Opponents Say It's Too Invasive," the paper quoted Barry Steinhardt of the American Civil Liberties Union as saying that the model act "gives governors and state health officials a blank check to impose the most draconian sorts of measures." It also quoted Andrew Schlafly, of the conservative Association of American Physicians and Surgeons, as saying that the model law "goes far beyond bioterrorism. . . . Unelected state officials can force treatment or vaccination of citizens against the advice of their doctors" (Hall 2002).

The legislation itself has created strange alliances and even stranger splits in the states. According to Tony Moulton, director of the Public Health Law Program of the Centers for Disease Control and Prevention (CDC), "In most states, ACLU chapters were opposed to the model act but others, such as Vermont, favored it. Some conservative administrators who helped write state legislation modeled on the act have been attacked by conservative organizations worried about increasing powers of the state." The broad concerns raised in the states that considered the model act were about civil liberties and private property rights. Questions were raised about whether the federal or state government can come in and take over a hospital. Does the state have the right to intrude on the doctor-patient relationship? Is forced quarantine (or even isolation) a real possibility? What should be the extent of compulsory government powers?

Public health administrators have been more receptive to the model act. Most practicing administrators in the various states said that their state did not engage in debate about the broad concerns among civil libertarians, academics, and constitutional lawyers regarding the model act's effect on civil liberties. Many administrators echoed the position of Robert Eadie, deputy director of the Nashville Metropolitan Health Department, who said that it was "nice to have guidelines for recommended best practice for emergency responses." These administrators also found it useful that the model act spells out the governor's responsibilities and the obligations and rights of public health personnel during an emergency.

One reason that many administrators were not taken aback by the scope of the model act's assertions of state power was the experience of the departments since the early 1980s with TB and HIV/AIDS. Departments of health, such as those in Tennessee, had been dealing with "noncompliant TB patients," Eadie said, and had been "very aggressive" in stemming the spread of TB in Nashville and Memphis. Their authority had evolved since the late 1990s, and the Tennessee legislature had a long tradition of working with civil liberties groups so the state had a basis for discussing the state health department's role in dealing with bioterrorism.

For some administrators, the model act, far from being a radical statement that could result in dramatic intrusions that undermined personal liberties, was itself a weak political compromise that did not give public health officials the tools that they need to adequately address the threat of bioterrorism or chemical attack. John Chapin, administrator of the Wisconsin Department of Health and Family Services, was chastened by his experience with the discussion of the model act in the Wisconsin legislature. "Wisconsin is a classic example of a state with relatively modern public health laws and strong laws," he observed. "A reasonable set of modifications to the law was proposed." But "it was then thrown into state politics and became a pawn in the partisan budget wars." The legislature responded not to true public health needs but to fear, he said.

Perhaps the aspect of the model law that has raised the most concerns is the provision for quarantine. George Annas has argued that there is "no empirical evidence that draconian provisions for quarantine, such as those outlined in the Model Act, are necessary or desirable" (2002, 1339). In some states it was very hard "selling the legislature that there was a need to look at quarantine laws," according to Minnesota representative Lee Greenfield. Legislators as well as the general population had forgotten that "previous epidemics required quarantine. That was history," Greenfield noted. In Connecticut and Ohio, civil liberties groups actively opposed the measures that gave public health authorities the right to quarantine, Ohio state health official Anne Harnish told us.

States dealt with these issues in a variety of ways. In Arizona the experience with TB had muted the arguments about quarantine, and those aspects of the model act that touched on quarantine were scaled down in light of people's antipathy toward intrusions by state authorities. Catherine Eden, director of the Arizona Department of Health Services, recalled that "there were things in there, like taking of people's property, confiscating weapons, getting onto people's property, which we thought would not be in the purview of public health, which were modified to make them more acceptable. . . . We put in provisions that people can't be forced to be vaccinated [or get] treatment but may be forced, if sick, to be isolated."

The other major civil liberties issue that caused some concern among state health officials was disease surveillance. The model act called for not only traditional disease reporting through hospitals and other health centers but for private physicians as well as pharmacists to reveal to state officials unusual disease patterns and specific patient information. Yet in few states did much real opposition to these new elements surface. For example, the state epidemiologist in Virginia collaborated with Johns Hopkins University to improve disease surveillance, and civil libertarians have voiced little concern about the issue. "Civil libertarians have their hands so full with guns and abortion that there is little time for concern about public health surveillance," Virginia's Kaplowitz told us. Most states appear to have been very careful to ensure that personal identifying information cannot be released, especially in light of the newly implemented privacy rule issued under the Health Insurance Portability and Accountability Act (HIPAA). This is the first federal privacy standard designed to protect the health records and other health information of health consumers, and it raised questions about the ethics of sharing information about communicable diseases with federal authorities, according to Daniel O'Brien, assistant attorney general for the Maryland Department of Health and Mental Hygiene (Moulton and Hodge 2003).

In sum, the issue that loomed largest for most public health personnel (as well as the broader public) was the apparent conflict between the growing power of the state and the threat to privacy and personal liberty. In light of the lack of public trust in the pronouncements coming from Washington—which appeared to many to overstate the threat of smallpox—much of the initial program to inoculate and to impose new reporting requirements was greeted with suspicion and passive rejection. Further, the state and federal perceptions of what was needed differed dramatically as state budget crises undermined popular public health programs and as federal mandates ignored the crises affecting the states' public health infrastructures.

Conclusion

It is clear from our interviews and reading of published and unpublished reports and media coverage that since September 11, much has been accomplished in terms of providing resources, legal reform, improved surveillance, and communication.

But, more generally, according to those we interviewed, it is unclear what the new attention to public and population health means for the long term. The early optimism regarding the new federal attention and funding waned as state legislators and public health officials faced state and federal budget crises and shifting federal attention in the war against terrorism. Budgetary crises in most states forced legislators to make hard choices about their priorities for education, social welfare, direct health services, and population health more generally. Even maintaining current levels of funding is inadequate to build up the public health infrastructure so that public health departments are prepared to counter terrorism and bioterrorism. Many public health officials were bitterly disappointed that federal attention to upgrading the public health infrastructure has waned in the years since September 11, 2001. But it is not unusual for the cause of the day to receive massive funding and for that funding to dissipate when new crises take center stage.

The maintenance of the public health infrastructure is probably the single most important means of preparing the nation for the myriad unpredictable tacks that outbreaks such as severe acute respiratory syndrome (SARS) and influenza can take. Without a strong permanent infrastructure of state health departments, the best emergency planning will be inadequate. The financial crises of the various states, combined with the shifting focus of the federal government, from bioterrorism and terrorism in general to smallpox and the war in Iraq more specifically, have lessened the early potential to enhance the system of services that is essential for improving the nation's efforts to address bioterrorism preparedness and the overall health needs of the American people.

Acknowledgments

The authors would like to acknowledge the generous support of a Robert Wood Johnson Investigators Award and a grant from the Milbank Memorial Fund.

References

Annas, George. 2002. Bioterrorism, public health and civil liberties. *New England Journal of Medicine* 346 (April 25): 1337–39.

Baird, J. Nick. 2003. Director, Ohio Department of Health. Interview by authors. October 28.

Bayer, Ronald, and James Colgrove. 2003. Rights and dangers: Bioterrorism and the ideologies of public health, pp. 51–74. In *In the wake of terror: Medicine and morality in a time of crisis,* edited by Jonathan D. Moreno. Cambridge, Mass.: MIT Press.

Bourne, Sandy Liddy, and Jennifer King. 2001. Issue alert to american legislative exchange council, environmental health attendees, re: model state emergency powers act. November 8. http://www.alec.org/viewpage.cfm?pgname=1.199 (accessed January 2003).

Chapin, John. 2003. Administrator, Wisconsin Department of Health and Family Services. Interview by authors. March 3.

Colmers, John M., and Daniel M. Fox. 2003. The politics of emergency health powers and the isolation of public health. *American Journal of Public Health* 93 (March): 397–99.

Coron, Angela. 2003. California Department of Health Services Director, Interview by authors, April 2.

Eadie, Robert B. 2003. Deputy director, Nashville Metropolitan Health Department. Interview by authors. March 21.

Eden, Catherine. 2003. Director, Arizona Department of Health Services. Interview by authors. March 21.

Gostin, Lawrence. 2000. *Public health law: Power, duty, restraint.* Berkeley: Milbank Memorial Fund and the University of California Press.

———. 2002. Public health law in an age of terrorism: Rethinking individual rights and common goods. *Health Affairs* 21 (November–December): 79.

Gostin, Lawrence O. et al. 2002. The model state emergency health powers act:

Planning for and Response to bioterrorism and naturally occurring infectious diseases. *Journal of the American Medical Association* 288 (August 7): 622–29.

Greenfield, Lee. 2003. Minnesota state representative. Interview by authors. April 16.

Hall, Mimi. 2002. Many states reject bioterrorism law; Opponents say it's too invasive. *USA Today*, July 23, p. A1.

Harnish, Anne. 2003. Assistant director, Ohio Department of Health. Interview by authors. April 1.

Kaplowitz, Lisa. 2003. Deputy commissioner for Emergency Preparedness and Response, Virginia Department of Health. Interview by authors. March 11.

McNeil, Donald G. Jr. 2003. After the war: Biological defenses; two programs to vaccinate for smallpox come to a halt. *New York Times*, June 19.

Moulton, Anthony. 2003. Director, Public Health Law Program, Centers for Disease Control and Prevention. Interview by authors. April 2.

Moulton, Anthony, and James G. Hodge Jr. 2003. Public Health Legal preparedness workshop summary, as of January 13, 2003. http://www.publichealthlaw.net/Resources/ResourcesPDFs/Dec11summ.pdf (accessed June 9, 2004).

Roser, Mary Ann. 2003. Smallpox: Is the U.S. vaccination program a shot in the arm for anti-terrorism or a shot in the dark by the White House? *Austin American Statesman*, January 19.

Rydell, Charlene. 2003. Health policy adviser to U.S. Rep. Tom Allen, D-Maine. Interview by authors. April 17.

Salinsky, Eileen. 2003. Public health emergency preparedness: Fundamentals of the "system." National Health Policy Forum background paper, Washington D.C., April 3.

19

Public Resistance
or Cooperation?

A Tale of Smallpox in Two Cities

Judith Walzer Leavitt

As we think about how to respond to current threats of bioterrorism and new emerging diseases, one major consideration must be the importance of gaining the trust and cooperation of the public. Public apprehension increases with each new disease since HIV/AIDS in the 1980s and, more recently, severe acute respiratory syndrome (SARS), monkey pox, Lyme disease, and West Nile virus. The potential for bioterrorists to use anthrax and other weaponizable biological threats makes the public much more edgy. In the face of decreasing trust in the ability of the public sector to be honest about any health problem that might emerge and to address it swiftly and effectively—distrust in government has been growing since the Vietnam War and Watergate—the potential for civic panic and social disorder in the face of a real public health emergency such as a reappearance of smallpox is significant. It behooves us to prepare ahead of time to try to avert such trauma to our cities and towns.

Looking at what happened when epidemics commonly attacked U.S. cities can provide some insight into strategies that might be worth adapting to today's situation and can also provide information about some things to avoid. While history cannot provide simple answers to our own debates, and actions cannot be merely transferred from the past to the present, a look back can teach us a lot. In this chapter, I concentrate on smallpox to make the bigger point about the importance of public health preparedness.

Historically, smallpox ravaged North American communities from the seventeenth century well into the twentieth century. The encounter with new European diseases often destroyed or decimated Native American communities, and the European settlers themselves periodically fell victim to this horrible disease.[1] Although vaccination provided a reliable preventative by the end of the eighteenth century, epidemics persisted throughout the nineteenth century and the first half of the twentieth century. The history of smallpox in the United States can be instructive in trying to understand why a disease with a clear preventative was not in fact prevented and in casting light on what factors are most important to success in public health campaigns against infectious diseases.

Two outbreaks of smallpox in two different U.S. cities are particularly relevant to our situation at the beginning of the twenty-first century. One occurred in Milwaukee, Wisconsin, which experienced a major smallpox outbreak in 1894 that led to a complete breakdown in civic order. In a context of political wrangling and medical dissension, the public resisted the health department mainstays of vaccination and isolation, which led to almost a month of rioting in the city streets, while smallpox spread widely and killed many who were exposed only because of the social disorder. The second smallpox example is the one that threatened New York City in 1947, and it illustrates the opposite response, civic order and citizen cooperation. New Yorkers stood in line for hours, full days, even came back the next day in some cases, waiting to get their vaccinations, and there was no sign of the kind of disturbances that characterized Milwaukee fifty-three years earlier. Smallpox did not gain a foothold in the city. How can we explain these diametrically different public responses, and what can we learn from them?

Smallpox in Milwaukee, 1894

Smallpox hit every ward in Milwaukee in June 1894, just months after the appointment of a new health commissioner, Walter Kempster, a British-born nationally recognized public health authority (Leavitt 1996). Immediately after taking office, Kempster tried to reorganize the health department, hire new people on the basis of merit rather than political patronage, even applicants from his own political party. With his natural allies, medical and political, angry with him, Kempster faced a challenging situation under the best of circumstances. Then, smallpox arrived and chaos erupted. As the health commissioner noted in the middle of the epidemic, "The alarm caused by a few cases of smallpox has served to un-balance the equanimity of the entire community" (Milwaukee Health Department Annual Report 1895, 20).

The health department response to the outbreak was the same as it had been during previous outbreaks. The agency started a vaccination campaign, opened the isolation hospital, and isolated some of the sick in their own homes, with placards posted outside declaring "Smallpox Within." Like his predecessors, Kempster usually used home quarantines in the middle- and upper-class areas of the city because he believed that people could be isolated at home more safely and effectively than in the hospital. In the poorer, more crowded immigrant sections of the city, he insisted that the ill be removed to the isolation hospital; in doing so he was acting under a city ordinance that gave him the power to do this by force if necessary (Leavitt 1996).

The health department's horse-drawn disinfecting van and the city ambulance became common sights on Milwaukee streets. Although the agency enforced a strict quarantine on those allowed to stay home, its dual policy was immediately viewed as discriminatory, especially because smallpox appeared in every ward in the city, and it led to criticism and re-sistance, especially because the health department was concentrating its efforts in the immigrant wards (Milwaukee Health Department Annual Report 1894, 1895; *Milwaukee Sentinel,* June 28 and July 4, 1894).

Some resistance to health department activities in 1894 focused on Kempster himself, in part because he represented a class and ethnicity distinct from the immigrants' and in part because he had disregarded pa-tronage in making his appointments (Leavitt 1996). The city (and the na-tion) was in the middle of an economic depression in 1894, and Kempster was not giving jobs to people who traditionally could have counted on political largesse.

But most of the resistance had nothing to do with Kempster and stemmed from distrust of vaccination itself. The Milwaukee Anti-Vaccination Society, the local branch of a national movement at the end of the nineteenth century whose aim was to prevent people from getting vaccinated, was very active in the city. Milwaukee had a strong contingent of antivaccinationists, people who believed that vaccination was either useless or injurious—a dangerous procedure (Milwaukee Anti-Vaccination Society 1891). These people, many of whom were immigrants, believed that vaccination did not protect against smallpox (they had cases to prove it) or that it spread other diseases (such as syphilis). The movement was based in the immigrant wards, but it had adherents around the city. It was fueled in part by lack of regulations concerning vaccination and repeated stories of unfortunate results of vaccination.

The medical community in Milwaukee was itself deeply divided on the subject of vaccinations, and probably one-third of Milwaukee physicians agreed with the antivaccinationists and objected to the procedure. Because the doctors disagreed, anyone who turned to a medical authority would have gotten various and mixed messages, and many doctors actively spoke out against the procedure altogether (Leavitt 1996).

The isolation hospital also was a point of contention in the summer of 1894. Despite renovations that had transformed the institution into what health officials insisted was a modern facility, residents of the neighborhood in which it was located still viewed it as a pesthouse and the source of much of their smallpox trouble. They claimed that it was a "menace to the health of citizens," a slaughterhouse where patients were "not treated like human beings" (*Milwaukee Sentinel,* July 23 and 24, 1894).

Another significant factor that led citizens to resist health department policies grew out of immigrants' fear of government authority. Officials in uniform who came knocking on the doors of private homes were specifically to be feared, especially when they tried to take children to the isolation hospital against their parents' will (Leavitt 1996). In Milwaukee, where the immigrant population was large, this fear surfaced repeatedly and was not adequately recognized by authorities as a factor that they needed to consider in determining health policy. One woman in a crowd of thousands explained that she would not let the city kill her child. "I can give [her] better care and nourishment here than they can give [her] in the hospital," she told a reporter for the *Milwaukee Sentinel* and *Evening Wisconsin* (see editions of August 6, 1894; Milwaukee Health Department Annual Report 1895, 42–43). The alderman representing the South Side district that became the focus of the smallpox outbreak used these sentiments to stir up support and protest.

As the riots erupted in early August, Alderman Robert Rudolph addressed the crowds with a speech that one city newspaper described as "not entirely free from incendiarism." Rudolph said, "I don't blame the people down here for being worked up. The patients at the hospital are not treated like human beings and the way the dead are buried is brutal" (*Milwaukee Sentinel,* August 8, 1894).

This perception, if not also the fact, of injustice in the way that government policy was enforced—that is, health officials allowed the rich to stay home and be quarantined, the poor were taken to a hospital said to treat patients badly—caused severe resentments. Those who lived on the South Side of Milwaukee believed that the rest of the city viewed them as the "scum of Milwaukee," as resident Henry P. Fisher told a reporter for the *Milwaukee Sentinel* (August 8, 1894), and they insisted on their rights. And this was but one example of how class and ethnic differences between the health department and city leaders on the one hand and the new immigrant settlers on the other affected the citizenry's response to the epidemic.

Milwaukeeans whose families were from Germany and Poland responded to the outbreak by not reporting cases of smallpox, by hiding the ill when health department people came to the door, and ultimately, that August, by rioting against the health department's methods of forcible removal, isolation, and vaccination (Leavitt 1996). Largely because of their coercive nature, the official policies—which seemed rational and essential to the health authorities—actually hastened the spread of the disease, insofar as they led directly to citizen hostility and resistance.

The newspapers followed the riots closely, characterizing them by the disruptions to city life that they caused and often emphasizing that, because police were reluctant to use their clubs on "feminine shoulders," women played a particularly large role. "Mobs of Pomeranian and Polish women armed with baseball bats, potato mashers, clubs, bed slats, salt and pepper and butcher knives lay in wait all day for the isolation hospital van," the *Milwaukee Sentinel* reported on August 30, 1894 (see also editions of August 12 and 30, and the *Milwaukee Daily News,* August 10, 1894). The rioters threw scalding water and pepper at the horses and assaulted the health officers with bats, clubs, and slats—domestic weapons in a civic battle.

The riots lasted about a month and made national news, reported in *Leslie's Illustrated Weekly* (figure 19.1), among other publications. The health department responded in a very dismissive, inflexible, and insensitive way, which exacerbated the problems. Kempster told a *Milwaukee Sentinel* reporter, "But for politics and bad beer, the matter would never have been heard of." He went on, insisting: "I am here to enforce the laws,

Figure 19.1 *Smallpox Troubles in Milwaukee,* from *Leslie's Weekly Illustrated Newspaper,* vol. 79, September 27, 1894, 207.

and I shall enforce them, if I have to break heads to do it. The question of the inhumanity of the laws I have nothing to do with" (August 7, 1894). The phrases "breaking heads" and "politics and bad beer" reverberated throughout the immigrant communities, reinforcing the view that the health department did not care about the poor.

As debates continued in the city council, and the riots continued on the streets, smallpox continued to claim lives. Because of the health department's coercive policies and the immigrants' resistance to them, unvaccinated Milwaukeeans continued to roam the streets, where they were exposed to smallpox (Leavitt 1996). Quarantines were impossible, and the spread of the disease therefore was inevitable. Although vaccinations were freely available, people refused them. During the violence on the South Side the daily work of the health department virtually came to a halt. Patients could not be removed to the hospital in the face of weapon-wielding mobs; patients from other parts of the city could not be transported to the hospital through the hostile eleventh ward. The health department was denounced for attempting to remove patients to the hospital and further censured when failure to remove such patients resulted in the spread of the epidemic. During the outbreak 1,079 Milwaukeeans contracted smallpox, and 244 died (Milwaukee Health Department Annual Report 1895).

The ethnic newspapers articulated the South Siders' position. None condoned the violence in the streets, but the German and Polish press understood, as Kempster never did, how the health commissioner was motivated by excessive legalism. He had not treated the sick with compassion and had allowed his agents to commit physical harm to people and to property in order to follow the letter of the law (see *Germania*, August 17; *Abendpost*, August 10; *Herold*, August 9, November 2; *Kuryer Polski*, August 8, 11, 13, 14, September 1, and October 19 and 31, all 1894).

The Milwaukee Common Council impeached Walter Kempster and threw him out of office for his controversial handling of the epidemic. After a year he was reinstated. But the impeachment cost the health department much of the authority that it had gained gradually during the nineteenth century, including the authority to forcibly remove anybody to an isolation hospital in the face of an outbreak.

New York City, 1947

The situation was very different in New York City in 1947. New York had not seen smallpox for more than a generation, and residents frequently ignored preventative programs like vaccination (Leavitt 1995). Only an estimated two million of the city's 7.5 million inhabitants had any degree of immunity. On March 1 Eugene Le Bar, a man from Maine on his way home from Mexico with his wife, got off a transcontinental bus in New York City because he was not feeling well. He checked into a midtown hotel for a rest, and when he was feeling better, the couple wandered around the Fifth Avenue shops. After five days, a rash erupted, and by then Le Bar was feeling so sick that he sought medical care at Bellevue Hospital. Because his rash was not familiar to hospital physicians, Le Bar was first admitted to the dermatology ward. His condition continued to deteriorate, however, and Bellevue's skin doctors transferred him to the isolation ward at Willard Parker Hospital with an unknown but suspected contagious condition. Le Bar died on March 10 of bronchitis with hemorrhages before a definite diagnosis of his malady could be made (what happened is reconstructed from the *New York Times* and *New York Daily Mirror*, April 5–30, 1947; see also Roueche 1965).

After two more cases of what some were beginning to recognize as smallpox, the staff and patients at Willard Parker were vaccinated. By late March health officials realized that, before he died, Le Bar had had many opportunities to expose people to smallpox in the streets of New York, in Bellevue Hospital, and among the nursing and medical staff at the Willard Parker Hospital. Two other early cases were traced to people who had been patients at Willard Parker when Le Bar was admitted, a child with tonsillitis and a man with mumps. Both were unvaccinated.

The public knew nothing of these early events. The health department did not reveal the problem until it received the laboratory reports on April 4. On April 5, New Yorkers opened their morning newspapers to learn of their risk of smallpox (Leavitt 1995). The health commissioner weighed his words to the press carefully. On the one hand, Health Commissioner Israel Weinstein assured the public that the chances of a full-scale epidemic were slight. He did not want to promote panic. On the other hand, Weinstein and Hospital Commissioner Edward M. Bernecker called for all who "have never been vaccinated or who have not been vaccinated since childhood to go at once to their doctors to receive this protection" (*Daily Mirror,* April 6, 1947). (For a good description of the relationship between the health department and the press, see Pretshold and Sulzer 1947.) The health department intended to give a strong message so that people would actually get vaccinated but not so strong that social order would break down.

The New York City Health Department's activities in 1947 differed greatly from what Milwaukee did fifty-three years earlier. New Yorkers proceeded on two fronts to prevent an epidemic from sweeping the city: First was the mass vaccination campaign; second was the case-tracing epidemiological work, carried out with the help of the U.S. Public Health Service. Health officers followed Le Bar's bus contacts around the country and contacted every person they could find who had been a patient at Bellevue and Willard Parker hospitals. Health officials instituted daily press conferences for the duration of the crisis, announcing all suspected and confirmed cases. Health officers understood the importance of keeping information flowing and did so with clear messages and frequent updates. They had the help of other city, state, and federal agencies, and, in particular, Mayor William O'Dwyer played an active role in coordinating events (Leavitt 1995).

Health officials posted signs and distributed lapel buttons everywhere:

"Be safe. Be sure. Get vaccinated." Weinstein or one of his deputies mounted regular radio shows about the diagnosis, nature of smallpox, and its spread—announcing details of every case—and worked hard to keep the public informed. New Yorkers understood that they were getting honesty and justice from their health department and other city officials, in large part because citizens felt that they were being informed of every detail as events unfolded.

The vaccination campaign starred free and voluntary vaccinations, available under health department auspices as well as through private physicians. Although the health department had the legal authority to forcibly remove people to the hospital and to demand vaccinations, it did not need to use that authority. The sick were all voluntarily hospitalized for their care. Given the enormous outpouring of public support and compliance, it was obvious that the vaccination campaign was successful without coercion.

The health department provided free vaccinations at thirteen hospitals, eighty-four police precincts, and every public and parochial school in New York City's five boroughs. Amid great publicity, O'Dwyer submitted to the vaccination procedure while urging all New Yorkers to follow his example. Each day the newspapers trumpeted the numbers vaccinated. In the first two weeks, five million New Yorkers were vaccinated, and by the fourth week, 6,350,000 New Yorkers had been vaccinated. People stood in lines that wound around city streets as they waited their turn to be vaccinated (figure 19.2). This was a large, unprecedented, and successful smallpox vaccination campaign, and it has been rightfully credited with saving the lives of perhaps hundreds of New Yorkers.

President Harry Truman submitted to a public vaccination and then visited New York. Federal cooperation in this outbreak was significant, especially around vaccine production (Leavitt 1995). New York relied on federal laboratories to produce and bring in vaccine. Public and private laboratories around the city were pushed into service. When drug companies hesitated to cooperate, O'Dwyer locked company representatives inside city hall and told them they were going to produce more vaccine and that they would make it available to the city at a reasonable price, or they would not be leaving the building. They agreed (Benison 1967, 385–88).

Men and women volunteers from the Red Cross, the Civilian Defense Volunteer Organization, teachers' groups, and women's clubs around the city worked in the vaccination program. Public compliance was high in part because in the immediate postwar years, people were accustomed to such shows of public cooperation and compliance. Wartime volunteers were still available, and the public retained something of an emergency mentality and readily followed government advice. The high level of citizen cooperation was related, too, to the level of organization in the city and the cooperative multijurisdictional effort. The strong health department infrastructure, built during the previous one hundred years, served the city well in this crisis. Statisticians calculated at the time that, based on the previous New York outbreak of smallpox, the city should expect more than four thousand cases and nine hundred deaths; the city reported only twelve cases and two deaths (Leavitt 1995).

The result was quite convincing and Health Commissioner Weinstein willingly spread the credit widely. He thanked the press and the radio, his staff, the nurses, city employees, and the mayor. Most of all, he credited the "intelligent cooperation of the public and the generosity of private physicians and volunteer workers," without whom, he said, "it would have been impossible to have achieved this remarkable record" (Weinstein 1947, 1381; see also City of New York Department of Health 1949). He was right.

Comparing the Milwaukee and New York Experiences

There are many relevant comparisons to be made between the events in Milwaukee and New York (figure 19.3). Both Milwaukee and New York had strong health departments. Although infrastructure was a crucial component of launching a successful campaign against infectious disease, infrastructure alone was not sufficient. The Milwaukee health department (as well developed as it was) should not have tried to handle the crisis on its own, and the state health department, which did not step in because it thought the city was doing the right thing in focusing on vaccination and isolation, should have played a more active role. In contrast, cooperation between local, state, and federal health officials brought needed services and personnel together and worked well in New York.

Milwaukee used strong-arm tactics on its poorer citizens; New York saved such pressure only for the drug companies. The question of coercion was extremely important and worked against Milwaukee and in favor of New York. The laws were similar in both states, but New York chose not to use the force of law and instead emphasized the cooperative aspects of the vaccination program. This policy won New York officials more allies in

Figure 19.2   A long line of New Yorkers waiting peacefully for smallpox vaccinations, near the entrance to the Morrisania Hospital in the Bronx, New York, April 14, 1947. AP/Wide World Photos.

the end. Milwaukee used discriminatory policies, applying the laws most stringently in the immigrant communities; New York led a voluntary city-wide program without making geographical, ethnic, or class distinctions. Again, the perception of fairness and honesty brought trust in the health department, and trust led to public cooperation with the campaign.

The media played crucial roles in both cities. In Milwaukee, the daily press provided limited information and mixed and partisan messages to its citizens; New York kept its citizens fully informed with clear messages and directions. New Yorkers knew what to do and where to go, and they learned the benefits of vaccination without a high fear level generated by the media.

Volunteer groups provided not just the staffing to make a vaccination campaign in New York successful; they also provided important links to community, through which people more willingly accepted the health department's activities. New York made full use of various religious and

```
┌─────────────────────────────────────────────────────┐
│                  Comparison                          │
│    Milwaukee            New York City                │
│   · Strong HD          · Strong HD                   │
│   · State help withdrew · State/Fed cooperation      │
│   · Strong-arm tactics  · Information/Respect         │
│   · Discriminatory      · Even handed                 │
│   · Limited information, · Media blitz, clear         │
│     mixed messages        message                     │
│   · No citizen activity · Use of citizen groups       │
│   · Raging epidemic     · Confined outbreak           │
└─────────────────────────────────────────────────────┘
```

Figure 19.3 Comparison of small-pox Milwaukee and New York.

secular organizations to support its efforts. Milwaukee did no such thing and paid the price of not having easy communication with those residents it most needed to reach.

In the end, Milwaukee suffered a raging epidemic and New York prevented a major outbreak. Health department activities proved crucial to the outcomes, but the content of the health policy itself was only one part of New York's success. How the message was delivered and the city-wide context created for it proved the importance of political and social context to successful public health work. Public response matters: A positive reaction can lead to successful programs, a negative one can prolong and exacerbate the outbreak of infectious disease and put more people at risk.

Implications for Today

While it is not possible simply to bring these historical examples into the present and try to apply them (because context and time period played significant roles in what happened in Milwaukee and in New York), some implications are worth discussing today.

Historically, the public's responses to the crisis of smallpox have covered the spectrum from strong resistance and rioting to active cooperation and compliance. The actual message from the two health departments was similar and focused on the importance of isolation and vaccination in smallpox prevention. But the methods of the two departments differed greatly and determined, to a large degree, how people received and processed that message.

The examples of Milwaukee and New York City demonstrate the need for a well-supported public health structure that is integrated not just locally but at all levels of government. It must be in place and functioning well when the crisis occurs, so it can immediately begin to do its work. Both Milwaukee and New York had active and fully staffed health departments when smallpox appeared. Unfortunately, at the beginning of the twenty-first century, health departments in the United States cannot claim to be running at peak levels. As new diseases emerge today, and with the threat of bioterrorism, we need strong health department infrastructures (Leavitt and Leavitt 2003).

The experiences in Milwaukee and New York City show that when coercion (either for quarantine or for vaccination) has been tried in the past, it has not led to effective solutions but to increased problems and public resistance. As New York City demonstrated in 1947, voluntary programs, coupled with clear, frequent, and honest bulletins from health authorities, worked best in garnering citizen support. In Milwaukee in 1894, on the other hand, coercion by health authorities—using powers that the legislature had given them—led to strong resistance and civic disorder. The health commissioner was not wrong in calling for vaccination and isolation, but the coercive force and complete lack of cultural sensitivity behind the message led directly to the failure to gain citizen cooperation.

Another clear message from the historical experiences is that fostering cooperation among voluntary groups and adding a strong public education component guided by a central agency, in this case the health department, has been much more successful than trying to work only from the top down. The effectiveness of a network of community activity was demonstrated well in 1947 in New York City, when women's clubs and teachers' organizations joined with Red Cross volunteers, medical volunteers, and city agencies to bring about an orderly and successful vaccination program.

History has shown repeatedly that the public responds with support when it receives frequent and honest information and communication. The message from health officials needs to be clear and accurate, and it needs to be presented directly to the public, without being filtered through the media. The Milwaukee press was thoroughly partisan during the Milwaukee outbreak, lining up either for or against Kempster, whereas the New York media provided more of an outlet for information. The information that the media provided demonstrated respect for the public's need to know. This does not mean that journalists should be the voice of the government; the role for investigative and interpretive journalism remains large. But it is equally true that in times of crisis, people need to be fully and honestly informed if they are to trust their leadership.

Leaders themselves must respect the public's need to know and provide the kind of information that people need. The more they try to hide, the more suspect they will become. The Milwaukee health department tried to force its activities without demonstrating sufficiently why and how citizens should cooperate, and it acted more leniently with middle-class native-born citizens than it did with the immigrant groups. This is perhaps the most important historical lesson of all: An essential element of public cooperation is a perception (or, better, the reality) of justice and equity. If history is any judge, unless the public is convinced it is receiving fair treatment, equally applied, it will resist public health policy instead of supporting it.

Smallpox historically is only one of many infectious diseases that have struck the human race. Although it seemed during the middle of the twentieth century that antibiotics and modern medicine had conquered such threats, it is abundantly clear at the beginning of the twenty-first century that emerging new infections will continue to affect global populations for the foreseeable future. Looking back at effective and ineffective ways of gaining public cooperation for public health activities through the example of smallpox can help us plan for the battles against disease that undoubtedly will be a part of our future.

Notes

This chapter is based on Judith Walzer Leavitt's article of the same title in *Biosecurity and Bioterrorism: Biodefense Strategy Practice and Science* 1, no. 3 (2003): 185–92, and is used here with permission.

1. See chapter 16 for a helpful description of how the smallpox disease pro-

gresses. For an excellent account of smallpox in the eighteenth century, see Fenn 2001.

References

Benison, Saul. 1967. *Tom Rivers: Reflections on a life in medicine and science.* Cambridge, Mass.: MIT Press.

City of New York Department of Health. 1949. *Report of the department of health city of New York for the years 1941–1948.* New York: Department of Health.

Fenn, Elizabeth. 2001. *Pox Americana: The great smallpox epidemic of 1775–82.* New York: Hill and Wang.

Leavitt, Judith Walzer. 1995. Be safe. Be sure: New York City's experience with epidemic smallpox, pp. 95–114. In *Hives of sickness: Public health and epidemics in New York City,* edited by David Rosner. New Brunswick, N.J.: Rutgers University Press.

————. 1996. *The healthiest city: Milwaukee and the politics of health reform.* Madison: University of Wisconsin Press.

Leavitt, Judith Walzer, and Lewis A. Leavitt. 2003. After SARS: Fear and its uses. *Dissent,* Fall. pp. 54–58.

Milwaukee Anti-Vaccination Society. 1891. Quotations from eminent medical authorities showing the utter uselessness and injuriousness of vaccination. Pamphlet. May. (A copy is available at the Middleton Health Sciences Library, University of Wisconsin, Madison.)

Pretshold, Karl, and Carolyn C. Sulzer. 1947. Speed, action, and candor: The public relations story of New York's smallpox emergency. *Channels* 25 (September 1947): 3–6.

Roueche, Berton. 1965. A man from Mexico, pp. 91–108. In *Eleven blue men.* New York: Berkley Medallion.

Weinstein, Israel. 1947. An outbreak of smallpox in New York City. *American Journal of Public Health* 37:1381.

Contributors

TAMAR BARLAM, M.D. received her degree from the University of Rochester School of Medicine and received her infectious disease training at the Columbia-Presbyterian Medical Center. She served on the infectious disease faculty at St. Luke's–Roosevelt Hospital Center in New York and the Beth Israel Deaconess Medical Center in Boston. In addition, she was an assistant clinical professor in medicine at the Columbia College of Physicians and Surgeons and assistant professor of medicine at the Harvard Medical School. She has a long-standing interest in the appropriate use of antibiotic drugs and in 2000 joined the Center for Science in the Public Interest in Washington, D.C., to direct the Project on Antibiotic Resistance. Currently, she is on the infectious disease faculty at the Boston Medical Center and the Veterans Administration Boston Healthcare System and serves as associate professor of medicine at the Boston University School of Medicine.

PAUL R. BILLINGS, M.D., PH.D., is an internist and medical geneticist who is also an adjunct professor of anthropology at the University of California, Berkeley. He chairs the Council for Responsible Genetics and is vice president and national director for genetics and genomics at Laboratory Corporation of America Holdings.

BERKELEY BIOTECHNOLOGY WORKING GROUP members share and critique research on the ecological, economic, political, social, and ethical contexts and effects of biotechnology. Along with Paul R. Billings, Richard C. Strohman, and Kenneth A. Worthy, contributing authors to

327

this chapter include Jason A. Delborne, a doctoral candidate at the University of California, Berkeley, who is researching the management of scientific dissent in agricultural biotechnology; Earth Duarte-Trattner, also at Berkeley, where he is completing his dissertation, "From Slaves to Transgenic Organisms: A History of Branding in Mexico and California"; Nathan Gove, who holds a master's in environmental science, policy, and management from Berkeley and currently works at the Biological Resources Research Center, University of Nevada, Reno; Daniel R. Latham, a doctoral candidate at Berkeley who is researching the biological control of insect pests in organic fruit orchards; and Carol J. Manahan, a student at the Graduate Theological Union, Berkeley, who is writing her dissertation, entitled, "The Moral Economy of Corn: StarLink and the Ethics of Resistance." Delborne and Manahan founded the group in 2001, inspired by conversations with Professor Charles Weiner.

DAVID L. DEMETS is professor and chair of the Department of Biostatistics and Medical Informatics at the University of Wisconsin–Madison. He received his doctoral degree in 1970 from the University of Minnesota, spent twelve years at the National Institutes of Health, and has been at the University of Wisconsin since 1982. DeMets is a recognized international leader in statistical research and methods for the analysis of clinical trials. He has coauthored nearly two hundred papers and book chapters, as well as two books, *Fundamentals of Clinical Trials* and *Data Monitoring Committees in Clinical Trials: A Practical Perspective.*

PAUL GEPTS is professor in the Department of Agronomy and Range Science at the University of California, Davis. His research focuses on the study of evolutionary and molecular mechanisms responsible for crop biodiversity, including the process of domestication, with emphasis on Phaseolus beans and the cowpea. He has conducted fieldwork in Latin America and Africa, as well as laboratory work in the United States. He is a Fellow of the American Association for the Advancement of Science (2001) and the American Society of Agronomy (2003).

LAWRENCE GOSTIN is professor of law at Georgetown University; professor of Public Health at the Johns Hopkins University; and director of the Center for Law & the Public's Health at Johns Hopkins and Georgetown Universities (CDC Collaborating Center "Promoting Public Health through Law"; http://www.publichealthlaw.net). He is a research fellow at the Centre for Socio-Legal Studies, Oxford University. Professor Gostin has led major law reform initiatives for the U.S. government, including the Model State Emergency Health Powers Act (MSEHPA) to combat bioterrorism and other emerging health threats. Gostin's latest books are *The AIDS Pandemic: Complacency Injustice, and Unfulfilled Expectations* (2004); *The Human Rights of Persons with Intellectual Disabilities: Different But Equal* (2003); *Public Health Law and Ethics: A Reader* (2002); *Pubic Health Law: Power, Duty, Restraint* (2000).

JO HANDELSMAN is Howard Hughes Medical Institute Professor in the Department of Plant Pathology and codirector of the Women in Science and Engineering Leadership Institute at the University of Wisconsin–Madison. Her work has appeared in a wide array of scientific journals, and she is co-author of *Biology Brought to Life: A Guide to Teaching Students to Think Like Scientists* (2002).

JAMES G. HODGE JR. is a lawyer and assistant public health professor at Johns Hopkins Bloomberg School of Public Health; the executive director for the Center for the Law and the Public's Health; a core faculty member of the Berman Bioethics Institute at Johns Hopkins Bloomberg School of Public Health; and an adjunct professor of law at Georgetown University Law Center.

JUDITH A. HOUCK is an assistant professor in the Departments of Medical History and Bioethics, Women's Studies, and History of Science at the University of Wisconsin–Madison. Her research focuses on the history of women's health. She is finishing a book on the history of menopause in United States. Her next project examines the twentieth-century women's health movement.

ABBY J. KINCHY is a graduate student in the Departments of Sociology and Rural Sociology at the University of Wisconsin–Madison. Her research focuses on problems of power and expertise in struggles about agriculture, biotechnology, and international trade. Her recent publications include an article on the Ecological Society of America in *Social Studies of Science* as well as work with Daniel Kleinman on policy debates about recombinant bovine growth hormone in *Sociological Quarterly* and

Science as Culture. In 2003 she received the Department of Rural Sociology's John H. Kolb Award for academic achievement.

DANIEL LEE KLEINMAN is associate professor in the Department of Rural Sociology at the University of Wisconsin–Madison, where he is affiliated with the Holtz Center for Science and Technology Studies and the Integrated Liberal Studies Program. He is the author of *Impure Cultures: University Biology and the World of Commerce* (2003) and *Politics on the Endless Frontier: Postwar Research Policy in the United States* (1995). He is the editor of *Science, Technology and Democracy* (2000).

JUDITH WALZER LEAVITT is the Rupple Bascom and Ruth Bleier WARF Professor of Medical History, History of Science, and Women's Studies at the University of Wisconsin–Madison, where she has taught since 1975. She has written or edited seven books, including *Typhoid Mary: Captive to the Public's Health* (1996), *Brought to Bed: Childbearing in America, 1750–1950* (1986), *The Healthiest City: Milwaukee and the Politics of Health Reform* (1982, 1998), *Women and Health in America: Historical Readings* (1999), and *Sickness and Health in America: Readings in the History of Medicine and Public Health* (1997), the last coedited with Ronald L. Numbers. Leavitt chaired the history of medicine department from 1981 until 1993, was the Evjue-Bascom Professor of Women's Studies from 1990 through 1995, and served as associate dean for faculty in the medical school from 1996 to 1999. A recipient of the Arthur Viseltear Award from the American Public Health Association and the University of Wisconsin's Folkert Belzer Lifetime Achievement Award, she is a fellow of the American Academy of Arts and Sciences and has recently completed a two-year term as president of the American Association for the History of Medicine.

MARGARET LOCK is the Marjorie Bronfman Professor in Social Studies in Medicine and is affiliated with the Department of Social Studies of Medicine and the Department of Anthropology at McGill University. She is an officer of l'ordre national du Québec, a fellow of the Royal Society of Canada, was awarded the Prix Du Québec, domaine Sciences Humaines, in 1997, and in 2002 received the Canada Council for the Arts Molson Prize. Lock's monographs include *Encounters with Aging: Mythologies of Menopause in Japan and North America* (1993) and *Twice Dead: Organ*

Transplants and the Reinvention of Death (2002), both of which received numerous prizes. She has edited or coedited ten other books and written more than 160 scholarly articles.

GERALD MARKOWITZ is adjunct professor in the Department of Sociomedical Sciences at the Mailman School of Public Health, as well as Distinguished Professor of History at John Jay College of Criminal Justice and the CUNY Graduate Center. He has been awarded numerous grants, including those from the National Endowment for the Humanities and was a recipient of the Viseltear Prize from the Medical Care Section of the American Public Health Association for "Outstanding Contributions" to the history of public health. With David Rosner, he has coauthored and edited many books and articles, including *Deceit and Denial: The Deadly Politics of Industrial Pollution* (2002), *Deadly Dust: Silicosis and the Politics of Occupational Disease in Twentieth Century America* (1991; 1994), *Children, Race, and Power: Kenneth and Mamie Clark's Northside Center* (1996), *Dying for Work* (1987), and *"Slaves of the Depression," Workers' Letters about Life on the Job* (1987).

BRIAN MARTIN is associate professor of science, technology, and society at the University of Wollongong, Australia. He worked as an applied mathematician before moving to social science. He has studied controversies involving nuclear power, pesticides, fluoridation, nuclear winter, repetitive strain injury, and the origin of AIDS, in addition to doing research on dissent, nonviolence, democracy, and other topics. He is the author of ten books and numerous articles and is international director of Whistleblowers Australia.

CHRISTINE MLOT is a science writer based in Madison, Wisconsin, and teaches writing at the University of Wisconsin. She has written for *Science, Science News,* and numerous other publications. Her chapter on antibiotic resistance and agriculture in this book is based on a feature article she wrote for *BioScience.*

R. DENNIS OLSON is the director of the Trade and Agriculture Project at the Institute for Agriculture and Trade Policy (IATP) in Minneapolis, working on agricultural, trade, and biotechnology issues. In 2003, Olson organized programs for activists and trade ministers on agricultural and trade policies both at the World Trade Organization negotiations in Cancun, Mexico, and the Free Trade Area of the Americans negotiations in Miami. Before coming to IATP, he worked as a community organizer for seventeen years with grassroots organizations in North Dakota and Montana on agricultural, environmental, and trade issues. He was graduated from the University of Montana in 1983 with a combined bachelor's degree in history and political science and a minor in Russian.

PETER H. RAVEN is director of the Missouri Botanical Garden and Engelmann Professor of Botany at Washington University in St. Louis. During his tenure there, he has built the Garden's international research program and become a worldwide advocate of conservation and sustainable development. His jointly authored botany text, *The Biology of Plants,* has been the standard in the field for more than thirty years. Raven, current president of Sigma Xi, has won numerous honorary degrees, medals, and awards for his efforts, including the National Medal of Science.

DAVID ROSNER is professor of History and Public Health at Columbia University and director of the Center for the History of Public Health at Columbia's Mailman School of Public Health. In addition to receiving numerous grants, he has been a Guggenheim fellow, a National Endowment for the Humanities fellow and a Josiah Macy fellow. Presently, he is the recipient of a Robert Wood Johnson Investigator Award. He has been awarded the Distinguished Scholar's Prize from the City University and recently the Viseltear Prize for Outstanding Work in the History of Public Health from the APHA and the Distinguished Alumnus Award from the University of Massachusetts. He is author of *A Once Charitable Enterprise*(1982, 1987) and editor of *"Hives of Sickness": Epidemics and Public Health in New York City* (1995) and *Health Care in America: Essays in Social History* (with Susan Reverby). He and Gerald Markowitz have recently authored *Deceit and Denial: The Deadly Politics of Industrial Pollution,* (2002).

ABIGAIL A. SALYERS is professor of microbiology at the University of Illinois. She is the author of numerous articles on antibiotic resistant bacteria and of two textbooks on microbiology. She was president of the American Society for Microbiology in 2000–2001, when she helped to develop the society's policy statement on agricultural use of antibiotics. She is a member of an advisory committee that considers antibiotic use issues for the Food and Drug Administration.

In 1960, BARBARA SEAMAN introduced a style of health reporting that focused more on the patient and less on the medical fads of the day. Her books include, among others, *The Doctors' Case Against the Pill* (1969), *Free and Female* (1972), and *The Greatest Experiment Ever Performed on Women* (2003). Seaman is cofounder of the National Women's Health Network and is a national judge of the annual Project Censored Awards. In February 2004, her highly controversial article on the lack of informed consent by fertility clinics was a cover story in *O, the Oprah Magazine*.

RANDALL S. SINGER is assistant professor of epidemiology in the Department of Veterinary and Biomedical Sciences and in the School of Public Health at the University of Minnesota. His research focuses on the transmission of infectious agents and combines empirical field and laboratory studies with simulation modeling approaches. Recently, he has focused on the area of international public health. In 2000 he received the Presidential Early Career Award for Scientists and Engineers from the United States Department of Agriculture.

ALLISON A. SNOW is professor in the Department of Evolution, Ecology, and Organismal Biology at Ohio State University. Her research focuses on environmental effects of genetically engineered crops, especially with regard to gene flow. She is the lead author of the Ecological Society of America's position paper on genetically engineered organisms and is currently president of the Botanical Society of America. She recently contributed to two biotechnology reports by the National Academy of Sciences Research Council. In 2002, Snow received a science and technology leadership award from *Scientific American*.

LESLEY STONE is a senior fellow in law and public health at the Center for Law and the Public's Health at Georgetown and Johns Hopkins Universities. Her research focuses on bioterrorism, AIDS, human rights, and the built environment.

RICHARD C. STROHMAN is professor emeritus of molecular and cell biology at the University of California, Berkeley. His research career has been devoted to fundamental questions of cell and tissue growth regulation, and cellular differentiation using molecular and cell approaches. A frequent contributor to the journal *Nature Biotechnology,* Strohman continues to be an active participant in global debates about the growing crisis in theoretical biology, holding the view that genetic determinism, the major component of biological reductionism, is increasingly unable to contend with newer findings of biological complexity and that a new, more holistic scientific theory of living systems is required—the subject of his forthcoming book. He chaired the zoology department at the University of California, Berkeley (1973–76) and served as director of Berkeley's Health and Medical Sciences Program (1976–79), as well as research director in 1990 for the Muscular Dystrophy Association's international effort to combat genetic neuromuscular diseases.

SYLVIA WASSERTHEIL-SMOLLER is professor in the Department of Epidemiology and Population Health, head of the Division of Epidemiology, and holds the Dorothy Manealoff and Molly Rosen Chair in Social Medicine at the Albert Einstein College of Medicine at Yeshiva University. She is the author of *Biostatistics and Epidemiology, A Primer for Health and Biomedical Professionals* (2004). Smoller's research has spanned both cancer and cardiovascular disease, and she has brought her expertise in both areas to bear as the principal investigator of the Women's Health Initiative.

DIXIE D. WHITT is a faculty member at the University of Illinois College of Medicine, where she teaches microbiology to first- and second-year medical students. She is the coauthor of two textbooks, *Bacterial Pathogenesis: A Molecular Approach* and *Microbiology: Diversity, Disease, and the Environment* and is working on a book on antibiotic resistance. In 2004 she received the Special Tribute Award from the graduating class of the College of Medicine. She chairs the membership committees of the American Society for Microbiology and the International Association of Medical Science Educators.

KENNETH A. WORTHY is a doctoral candidate in environmental science, policy, and management at the University of California, Berkeley, where he specializes in environmental philosophy and ethics. His dissertation examines how structural characteristics of modernity play a role in environmental degradation.

Index